Physically Unclonable Functions

Basel Halak

Physically Unclonable Functions

From Basic Design Principles
to Advanced Hardware Security
Applications

 Springer

Basel Halak
Southampton
UK

ISBN 978-3-030-08293-2   ISBN 978-3-319-76804-5   (eBook)
https://doi.org/10.1007/978-3-319-76804-5

This Springer imprint is published by the registered company Springer International Publishing AG
part of Springer Nature
The registered company address is: Gewerbestrasse 11, 6330 Cham, Switzerland

To Suzanne, Hanin and Sophia
as well as to my parents

Preface

Computing devices are increasingly forming an integral part of our daily life; this trend is driven by the proliferation of the Internet of things (IoT) technology. By 2025, it is anticipated that the IoT paradigm will encompass approximately 25 billion connected devices. The interconnection of such systems provides the ability to collect huge amounts of data which are then processed and analysed for further useful actions. Applications of this technology include personal health monitoring devices, the smart home appliances, smartphones, environmental monitoring systems and critical infrastructure (e.g. power grids, transportation system and water pipes).

The pervasive nature of this technology means we will soon be finding computing devices everywhere around us, in our factories, homes, cars and even in our bodies in the form of medical implants. A significant proportion of these devices will be storing sensitive data. Therefore, it is very important to ensure that such devices are safe and trustworthy, however; this is not an easy task, and there still many challenges ahead.

First, there is a rising risk of hardware Trojans insertion because of the international and distributed nature of the integrated circuits production business. Second, security attacks, especially those that require physical access to the device under attack, are becoming more feasible given the pervasive nature of IoT technology. Third, most of IoT devices are considered to be resource-constrained systems, which makes it prohibitively expensive to implement classic cryptographic algorithms; in those cases, a cheaper and more energy-efficient solution is required.

Physically unclonable functions (PUFs) are a class of novel hardware security primitives that promise a paradigm shift in many security applications; their relatively simple architectures can answer many of the above security challenges. These functions are constructed to exploit the intrinsic variations in the integrated circuit fabrication process in order to give each silicon chip a unique identifier, in other words, a hardware-based fingerprint.

The versatile security applications of the PUF technology mean that an increasing number of people must understand how it works and how it can be used in practice. This book addresses this issue by providing a comprehensive introduction on the design, evaluation metrics and security applications of physically unclonable functions. It is written to be easily accessible by both students and engineering practitioners.

Chapter 1 of this book gives a summary of the existing security attacks and explains the principles of the cryptographic primitives used as the building blocks of security defence mechanisms; it then introduces the physically unclonable function (PUF) technology and outlines its applications. Chapter 2 explains the origin of physical disorder in integrated circuits and explains how this phenomenon can be used to construct different architectures of silicon-based PUFs; it also outlines the metrics used to evaluate a PUF design and gives the reader an insight into the design and implementation of PUF on configurable hardware platforms. Chapter 3 explains the physical origins of the major reliability issues affecting CMOS technology and discusses how these issues can affect the usability of PUF technology; it also presents a case study on the evaluation of the impact of CMOS ageing on the quality metrics of PUF designs. Chapter 4 provides a comprehensive tutorial on the design and implementation principles of error corrections schemes typically used for reliable PUF designs; it also explains in details the state-of-the-art pre-reliability enhancement processing approaches applied at the chip post-fabrication stage. Chapter 5 investigates the security of PUF as cryptographic primitives; it discusses the existing attacks on PUFs and possible countermeasures and, in addition, introduces a number of quality metrics to evaluate the security of PUF designs. Chapter 6 focuses primarily on how to use PUF technology in practice; it explains in detail how PUF technology can be used to securely generate and store cryptographic keys, construct hardware-assisted security protocols, design low-cost secure sensor, develop anti-counterfeit solutions and implement anti-tempter integrated circuits.

This book has many features that make it a unique source for students, engineers and educators. The topics are introduced in accessible manner and supported with many examples. The mathematics are explained in detail to make them easy to understand. Detailed circuit diagrams are provided for many of the PUF architectures to allow reproducibility of the materials. Each chapter includes a list of worked exercises, which can be an excellent resource for classroom teaching. In addition, a list of problems is provided at the end of each part; in total, this book contains more than 80 problems and worked examples. The appendices include exemplar digital design of PUF written in system Verilog and a list of MATLAB scripts used in this book to characterise PUF quality metrics. The detailed examples of PUF applications in Chap. 6 can be an excellent source for course projects. Each chapter ends with a conclusion, which summarises the important lessons and outlines the outstanding research problems in the related areas. The book has a large number of references which give plenty of materials for further reading.

How to Use this Book

The material of this book has evolved over many years of working on this topic in research and teaching. From our experience, one can teach most of the book contents in a one semester course, which includes 36 one-hour sessions. Some of the book chapters can be also taught separately as short courses or part of other modules. Here are a couple of examples

Short Course Example 1:

Title: The Principles of PUF Technology
Contents: Chapters: 1 & 2 and Appendices 1 & 2

Short Course Example 2:

Title: PUF-Assisted Security Protocols
Contents: Chapter 2, Sections 6.4 and 6.5 and Appendices 1 & 2

Graduate students and engineer, it is advisable that you read the book chronologically to achieve the best learning experience.
As an engineer, a researcher and an educator, I have worked in the field of hardware security for more than 10 years and was always fascinated by its many challenges and intriguing problems.

I hope you enjoy reading this book as much as I enjoyed writing it.

Southampton, UK   Basel Halak
January 2018

Acknowledgements

Very few books are entirely the work of one person's unaided efforts and this is no exception. I would like first to thank all those who wrote papers on physically unclonable functions which have helped me to develop my own understanding. The sheer fascination of this technology, some of these writers conveyed and their convictions in its versatile applications, was what attracted me to this topic in the first place. Second, I would like to thank my wife for her unwavering support, without which it could not be possible to complete this work. I am also grateful for my students for their inspiring questions and feedback comments that helped me to improve the text. Finally, special thanks to my colleagues at the University of Southampton for the informative discussions we had over the years.

Contents

About the Author

Dr. Basel Halak is the Director of the Embedded Systems Master program at Southampton University. He is a member of the Sustainable Electronics Research group, as well as Cyber Security group at Electronics and Computer Science School (ECS). He has written over 60 conference and journal papers, and authored two books. He has received his Ph.D. degree in Microelectronics System Design from Newcastle University. He was then awarded a knowledge transfer fellowship to develop secure and energy-efficient design for portable healthcare monitoring systems. His background is on the design and implementation of microelectronics systems, with special focus on developing secure hardware implementation for cryptographic primitives such as physically unclonable functions. He lectures on digital design, Secure Hardware and Cryptography, supervises a number of M.Sc. and Ph.D. students, and is also leading the European Masters in Embedded Computing Systems (EMECS). He is the recipient of the Vice Chancellor Teaching Award in 2016, and the bronze leaf award in IEEE PRIME conference for his paper on current-based physically unclonable functions. He is a senior fellow of the Higher Education Academy (HEA), a guest editor of the IET CDT, and serves on several technical program committees such as IEEE ICCCA, ICCCS, MTV, IVSW, MicDAT and EWME. He is also member of hardware security working group of the World Wide Web Consortium (W3C).

A Primer on Cryptographic Primitives and Security Attacks

1

1.1 Introduction

It was 2 a.m. in the morning, when Michael and Jane heard a man's voice, 'Wake up baby, Wake up…'. The couple run to their 2-year-old' daughter room. The baby was fast asleep, but the web-enabled baby monitor was moving; this was a device Michael and Jane had bought a year earlier to make it easier for them to check up on their daughter using an app on their smartphones. The couple then discovered that a hacker had managed to gain control of the device and was using it to watch and harass their child. One year later, a mother from New York City was horrified when she discovered by accident that a hacker was spying on her 3-year-old child using a similar device. One can only imagine the parents' shock when faced with such an unfortunate incident. These were not rare events; in fact, in the United States alone there were more than seven similar stories reported in the media between 2013 and 2015.

A more serious security breach took place in 2015, when a pair of hackers demonstrated how they could remotely hack into the 2014 Jeep Cherokee; what is more, they showed how they could take control of the vehicle and completely paralyse it whilst driven on the motorway. Later on, during the Black Hat conference the two researchers (Charlie Miller and Chris Valasek) explained in details how they had achieved this. They started by hacking into the multimedia system of the Jeep through the Wi-Fi connection; it emerged that it is not very difficult to do this because the Wi-Fi password is generated based on the time when the vehicle and its multimedia system are turned on for the first time. Although such approach of password generation is deemed to be secure, a hacker who manages to know the year the car was manufactured and guess the month it was first used can reduce the possible Wi-Fi passwords to around 15 million combinations, a small number from a hacker's perspective as it allows him to find the right password in few hours using a brute force search algorithm.

© Springer International Publishing AG, part of Springer Nature 2018
B. Halak, *Physically Unclonable Functions*,
https://doi.org/10.1007/978-3-319-76804-5_1

The above stories provide some useful insights into the current security challenges facing electronics systems' designers.

First of all, there is an increasing number of computing devices forming an integral part of our daily lives; this trend is driven by the proliferation of the Internet of things (IoT) technology, which is a network of physical objects connected to the Internet infrastructure to perform tasks without human interaction. These objects are typically embedded with electronics, software and sensors; this enables them to collect and exchange data. The network connectivity allows these devices to be operated and controlled remotely. The IoT technology is expected to be used in a wide variety of applications, such as personal health monitoring devices, smart home appliances, smart cars, environmental monitoring systems and critical infrastructure (e.g. power grids, transportation systems and water pipes) [1]. By 2020, it is anticipated that the IoT paradigm will include approximately 20 billion connected devices.

Second, there is an increasing reliance on portable hardware devices to carry out more security-sensitive tasks. The most prominent examples are the smartphones, enhanced with multitudes of sophisticated applications; these devices form an integral part of modern life. They are currently being used for socialising with friends (e.g. Facebook), finding a partner (e.g. Dating apps), shopping, streaming movies, gambling, carrying out bank transactions and doing business. In fact, we can safely assume that a mobile device contains more information on their owner than what their most intimate partner will ever know.

Furthermore, there are an increasing number of emerging applications which have stringent hardware security requirements. For example, mobile payment, biometric access control mechanisms and subscribed TV channels.

Designing secure systems, however, is not an easy task, and the complexity of computing device is increasing rapidly; in fact, current electronics systems can have more than one billion transistors. In order to ensure the security of such systems, one needs to verify the system is doing exactly what is supposed to do nothing more and nothing less; however, given the complex nature of modern designs and the large number of possible interactions between different components on the same chip (processors, memories, buses, etc.), validating the security of a device becomes an intractable problem.

On the other hand, there are many parties who stand to gain from weakly secured systems. For example, an adversary who manages to break the security of a smart card can potentially steal a large amount of money, and a customer who can gain unauthorised access to a box set to watch paid TV channels can save subscription fees. What is more, adversaries may have more sinister goals than accumulating wealth; for example, there is nothing preventing a resourceful adversary from crushing a weakly secured smart car and killing its passengers; he may not even be charged with the crime.

In summary, electronics systems are currently an integral part of modern life; therefore, it is vital to ensure that those devices are trustworthy.

The question is how to build computing devices we can trust?

To answer this question, a designer should first have a clear idea of the type and value of the assets they need to protect, and a very good understanding of the potential threats. This knowledge will allow him to develop appropriate defence mechanisms.

This chapter aims to:

(5)   Provide a summary of the existing security attacks;
(6)   Explain the existing cryptographic primitives used as the building blocks of security defence mechanisms;
(7)   Explain the forces driving the development of hardware-based security solutions;
(8)   Introduce the physically unclonable functions (PUFs) technology and outlines its applications.

It is hoped that this chapter will help the reader to develop a good understanding of the motivation of secure hardware design and how physically unclonable function fit in this context.

1.2   Chapter Overview

The remainder of this chapter is organised as follows. Section 1.3 provides a summary of known security attacks on electronics systems. Section 1.4 outlines the main cryptographic primitives typically used to develop defence mechanisms. Section 1.5 looks into the motivation behind the development of hardware-based security solutions. Section 1.6 introduces the concept of physically unclonable functions and briefly explains their main cryptographic applications. Conclusions are drawn in Sect. 1.7, followed by a list of exercises in Sect. 1.8.

1.3   An Overview of Security Attacks

In order to understand the scope and the nature of security attacks on electronics systems, we will consider a simple scenario that is applicable to a vast majority of applications. Envisage two communicating parties, Bob and Alice as shown in Fig. 1.1. Each of them uses an electronic device to send/receive messages over a communication link.

Envisage an adversary Eve who would like to spy on Bob and Alice; in principle, Eve can try to compromise one of the three components of such a system, namely, the communication link, the software running on Bob's or Alice's electronics devices or their respective hardware circuitry.

One can argue that Bob and Alice are also part of this system so they can also be a target for Eve's attack. Although this is true, this discussion is limited to the threat against electronics systems and does not include social engineering attacks.

The remainder of this section gives examples of the known security attacks on each of the above-stated components.

1.3.1   Communication Attacks

These are carried out by an eavesdropping adversary, who manages to gain unauthorised access to a communication channel; such access allows him to steal sensitive information or maliciously manipulate the transmitted data. Replay attacks are one of the examples of this type, wherein an attacker records a transmitted message and re-sends it at a later time; if such a message was sent from a customer to his bank and contained money transfer instructions, then a replay attack can lead to carrying out the same transactions multiple times without the consent of the customer.

1.3.2   Software Attacks

These attacks aim to maliciously modify software running on a computing device to install a malware and/or cause a system malfunction. An attacker can gain access to a computing device remotely if the latter is connected to the Internet or other public networks. Examples of remote access software attacks include the WannaCry ransom attack that hits tens of thousands of computer machines around the globe in 2017; in the United Kingdom alone, more than 40 NHS (National Health Service)

Fig. 1.1 An exemplar communication scenario

organisations were affected, which led to the cancellation of many non-urgent operations and GP appointments. This malware spreads exploiting a known vulnerability in Microsoft's Windows operating system.

If remote access is not feasible, attackers may try to secure a physical access to install the malware; one way to achieve this is by exploiting USB (Universal serial bus) devices. We know that a generic USB device indicates its capabilities through one or more descriptors (interface class code). The device could also de-register and register again as a different device with a different descriptor. This is all possible in the USB standard and it is not hard to see how this inherent trust on USB devices can break a system's security. The BadUSB research SRLabs [2] is an example of this, which shows that a hacker can reprogram a USB peripheral's firmware to change the behaviour of its microcontroller to perform malicious tasks if connected to a computer. Indeed, there are commercial USB-based keyloggers that record the user's keystrokes and transmit these wirelessly. Hackers introduced another USB attack with a cheap DIY hardware (TURNIPSCHOOL), hidden in a USB cable, which provides short-range RF communication capability to a computer. One of the most prominent examples of a security attack that was carried out using a malicious USB in recent history is the Stuxnet virus that attacked Iran's nuclear program; it is reported that the virus was planted by a double agent using a USB flash memory. The virus made its way into the equipment that controlled the centrifuges to enrich uranium. The malicious code is reported to cause the speed of centrifuge's rotor to increase and decrease which leads to excessive vibrations, ultimately breaking the centrifuge [3].

1.3.3   Hardware Attacks

These attacks aim to maliciously modify or tamper with the physical implementation of a computing device. They generally require direct access to the hardware or the design files, but can sometimes be carried out remotely as the case with the Stuxnet attack described above, wherein a manipulation of the software controlling the centrifuges led to physically damaging them; this is referred to as cyber-physical attacks.

In principles, hardware attacks can take place at any time during the life cycle of the computing device; based on this, they can be classified into two categories:

(a) **Attacks during design and fabrication**: An example of this type is Trojan insertion, which consists of adding extra circuitry that has malicious functionality into a design (e.g. a kill switch, a time bomb, etc.) [4]. Overproduction of integrated circuits is another example, wherein a malicious fabrication facility produces more chips than required, these are subsequently sold in the black market; this is referred to as IC counterfeiting [5].

(b) **Post-fabrication attacks** (also referred to as physical attacks) take place after the device is put in operation. This type can be further classified into three categories:

- Invasive attacks that require access to the internal structure of the integrated circuits: an example of this type is reverse engineering attacks which aim to steal the intellectual property of a design.
- Non-invasive physical attacks wherein an adversary interacts with the hardware externally: one example of this type is side-channel analysis wherein an attacker analyses the power consumption of the electromagnetic noise of a device in order to deduce sensitive information [6–8]; another example is data remanence, wherein one can retrieve information stored in a device memory even if they have been deleted [9].
- Semi-invasive attacks which require access to the surface of a device but not the internal logic: a prime example of this type is optical fault injection [10], wherein illuminating a device can cause some of its transistors to conduct current, which may trigger an error.

Hardware attacks typically require more resources and knowledge to wage which makes them less frequent compared to communication/software attacks; adversely, the cost of protecting electronics systems against this type is higher as illustrated in Fig. 1.2.

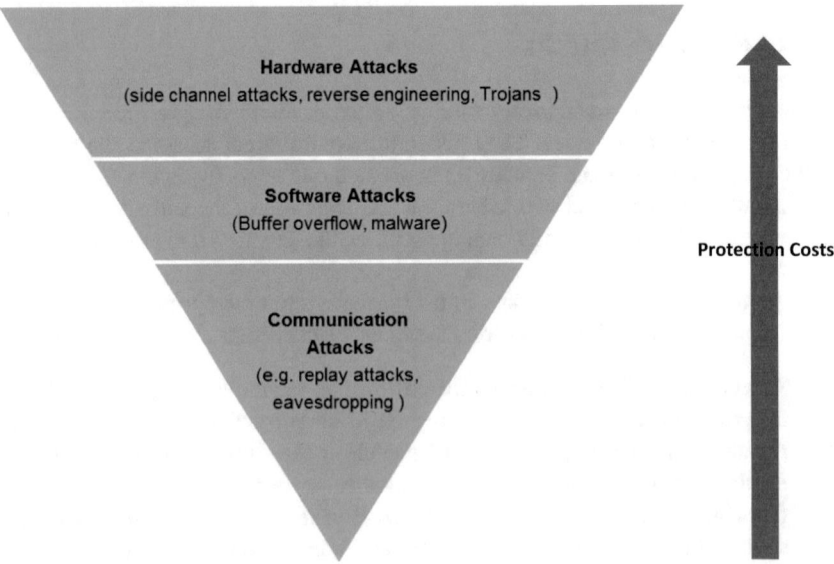

Fig. 1.2 Types of security attacks on electronics systems

1.4 Cryptographic Primitives

A cryptographic primitive can be defined as a low-level cryptographic algorithm which has a specific security-related functionality. These algorithms are the basic building blocks of secure systems; therefore, they must be carefully developed and rigorously checked. The functionality and security usages of the most widely used cryptographic primitives are explained below.

1.4.1 Symmetric Ciphers

A sympatric cipher can be defined as a deterministic algorithm that converts a stream of unencrypted data (aka plaintext) into a ciphertext; the latter has an illegible format that hides its information contents. The encryption and decryption processes rely on the use of a secret key that is only known to authorised parties. This means, prior to using symmetric ciphers, communicating parties need to agree on a shared key.

In principle, if an adversary manages to intercept the encrypted messages, they may still be able to decode them without having the decryption key (e.g. by using brute force search); however, encryption algorithms are designed and continuously revised to make such a possibility extremely unlikely. Indeed, the science of encryption has been the battleground for a raging war between cryptographers and cryptanalysts for more than 2000 years. The former group have been developing ever-stronger encryption algorithms and the second group are trying to find weakness in such algorithms.

The security of symmetric encryption algorithms is typically achieved by making the statistical relationships between the plaintext and the key on one hand and the ciphertext on the other hand as complex as possible; in addition, the key space is chosen very large such that finding the key using a brute force search is practically impossible.

Adversely, the technological advancement in the semiconductor industry made it possible to build high-speed computer machines that can be exploited by cryptanalysis; a prime example of this is the cracking of what was a 'well-established' encryption standard DES (data encryption standard) in the mid-nineties of the twentieth century [11]. This has triggered the development of a more secure algorithm known as advanced encryption standard (which is still being in use) [12].

Symmetric ciphers are building blocks for many cryptographic systems such as secure shell (SSH), a protocol used to operate networks services securely over unsecure communication channels. This makes services such as remote log in secure to use. They can also be used to encrypt on-chip communication links to protect against some types of physical attacks (e.g. probing internal signal on-chip and power analysis). Symmetric key ciphers are also used to construct message authentication codes used to ensure data integrity (i.e. the data has not been maliciously modified while being transmitted).

1.4.2   Asymmetric Ciphers

These are encryption algorithms, but unlike symmetric ciphers, they do not use the same key for the encryption and decryption processes; instead, each user has a public key that can be made available widely and a private key only known to them. If a sender 'Bob' wants to transmit an encrypted message to a receiver 'Alice', he encrypts his messages using Alice's public key, normally obtained from a trusted third party. Only Alice would be able to decipher Bob's messages using her private key.

The security of asymmetric cipher is based on the difficulty in solving a mathematical problem such as factorising a large integer number or computing discrete logarithms. Such problems are prohibitively complex to solve unless one has access to additional information, which is kept secret and used for decryption by authentic receivers (i.e. the private key).

One advantage of asymmetries ciphers compared to their symmetric counterparts is that they do not need a prior exchange of keys. On the other hand, these algorithms typically require more computation resources.

In practice, asymmetric ciphers are used to construct key exchange protocols to help communicating parties to agree on an encryption key; subsequently, a symmetric cipher is used for data encryption.

1.4.3   One-Way Hash Functions

This primitive maps arbitrary length inputs to a short fixed-length output (digest). A hash function should satisfy a number of requirements to be suitable for security applications. First of all, it should be deterministic (the same digest is obtained if the input is the same). Second, it should be extremely hard to regenerate a message from its hash value (i.e. digest). Third, it should be infeasible to find two messages which have the same hash value. Fourth, it should have a large avalanche effect, which means a small change in its inputs lead to a significant change on the output; this makes it harder for an adversary to build a correlation between the messages and their digests.

A classical application for this primitive is the secure storage of passwords file, wherein instead of storing the users' passwords as clear texts (i.e. unencrypted), which is a significant vulnerability, one can store the digest of each password; this makes it harder for an adversary to obtain the stored passwords if the server is compromised. In this case, to authenticate a user, a server recreates a hash value of the password presented at the time of authentication and then compares it with a previously stored hash. Keyed hash functions can be built using symmetric ciphers.

One-way hash functions are also used to construct messages authentication codes such as HMAC (Hash-based message authentication code) [13].

1.4.4 Random Number Generators

These primitives are mainly used to generate a nonce, which is an arbitrary number used only once. Nonce is employed as an initialization vector for encryption algorithms and in authentication protocols to prevent replay attacks.

1.4.5 Oblivious Transfer

An oblivious transfer (OT) protocol in its simplest form enables a sender to transfer one or multiple data items to a receiver while remaining oblivious to what pieces of information have been sent (if any). One form of this scheme is called 1-of-2 oblivious [14]; it allows one party (Bob) to retrieve one of the two possible pieces of information from another party (Alice), such that Bob does not gain any knowledge on the piece of data he has not retrieved nor Alice establishes which of the two data items she holds has been transferred to Bob. The 1-of-2 oblivious transfer has been later generalised to k-of-n OT [15], which can be used to construct secure multiparty computation schemes.

1.4.6 Bit Commitment

A commitment scheme is a cryptographic protocol which allows one party (referred to as the committer) to commit to a chosen value while keeping it secret from another party (referred to as the receiver). This is called the commitment phase. At a later stage called the reveal phase, the committer can prove his commitment to the receiver. This scheme has a number of important applications such as verifiable secret sharing [16] and secure billing protocols [17].

Security defence mechanisms (aka. countermeasures) can be built using a combination of the above primitives. For example, to develop secure electronic mail service, one can use the encryption primitive to ensure data privacy, message authentication codes to ensure data integrity and a hash function to protect the users' passwords stored on the mail server.

It is important when developing effective countermeasure to have a holistic view of the system in question and ensure each of its components satisfies the security requirements.

1.5 Why We Need Hardware Security?

The security of the physical layer (i.e. the Hardwar) of electronics systems has been gaining an increasing attention due to a number of issues summarised below:

(a) **The Rising Risk of Hardware Trojans**

The semiconductor chips' production chain is currently a multinational distributed business that relies heavily on the reuse of expertise and intellectual properties (IPs) from local and global sources; this increases the risk of an adversary injecting a malicious circuit into the design [5]. Nowadays, the first stage of developing a new chip includes outsourcing intellectual property (IP) designs from third-party design houses; the second stage takes place in-house and consists of system integration. In the third stage, a blueprint of the design (e.g. GDS-II layout format) is sent to a foundry (usually in a different country) that develops the fabrication masks and implements the design on silicon chips. The latter are tested at yet another geographical location in a third-party test facility. The distributed nature of IC production chain gives potential adversaries numerous opportunities to tamper with the original design to include malicious functionalities, for example, a hardware Trojan may be used to disable the chip remotely (i.e. kill switch) and create a backdoor or simply to leak information (e.g. parasitic antennas); there have been a number of recent reports which indicate that hardware Trojans are real threats [18, 19].

(b) **Physical Attacks are Becoming More Feasible**

This is due to the rise of Internet of things technology, which consists of connecting billions of computing devices to the Internet, such devices can include vehicles, home appliances, industrial sensors, body implants, baby monitors and literary any physical object which has sufficient capability to communicate over the Internet [20]. The pervasive nature of this technology makes it easier for an adversary to get hold of a device and carry out well-established physical attacks to extract sensitive data, inject a fault or reverse engineer its design.

(c) **The Increasing Need for Hardware -Assisted Cryptographic Protocols**

Conventional examples of secure hardware tokens include smart cards and staff cards. In recent years, these tokens are increasingly relied upon in cryptographic operations for a range of security-sensitive applications. For example, numerous European countries have implemented electronics identity systems used for services such as tax payment and retirement funds management. Examples of such devices are ID-porten in Norway and Telia ID in Sweden.

Another usage of secure hardware token is secure financial transactions; these days the majority of financial institutions offer their clients the ability to access their account online and carry out various tasks such as balance checking, money transfer and setting direct debit. To enhance the security of remote accounts management, extra measures are needed which go beyond the typical login/password approach. To meet such requirements, many banks have started giving their customers hardware devices, which are used to confirm their identities and sometimes to generate a digital signature to confirm the details of a transaction they are trying to make (amount, beneficiary name, etc.).

A third application area for secure hardware tokens is secure multiparty computation, wherein a number of parties need to carry out joint communication based on their private individual inputs; this type of computation is used in private data mining, electronic voting and anonymous transactions. Protocols that meet the security requirements of such applications do not typically have efficient implementation in practice [21]. This has given rise to hardware-assisted cryptographic protocols. The latter rely on the use of tamper-proof hardware tokens to help achieve strong security guarantees as set in the Canetti's universal composition (UC) framework [22]. In this type of protocols, the trust between communicating parties is established through the exchange of trusted hardware tokens.

1.6 Physically Unclonable Functions

The fabrication processes of physical objects can sometimes have some limitations, which make it difficult to have exact control of the devices being manufactured; this leads to slight variations in the dimensions of the resulting products. This limitation is especially true in the fabrication process of semiconductor devices in advanced technology nodes (below 90 nm) [23–25]. These intrinsic process variations lead to fluctuations in the transistors devices' length, width, oxide thicknesses and doping levels. This makes it impossible to create two devices which are identical. This means the same transistor fabricated on different devices may have slightly different electrical characteristics.

A physically unclonable function (PUF) exploits these inherent process variations to generate a unique identifier for each hardware device.

A PUF can be defined as a physical entity whose behaviour is a function of its structure and the intrinsic variation of its manufacturing process. This means two PUF devices will have two distinct input/output behaviours even if they have identical structures because of process variations [26]. PUFs can be realised using integrated circuits, in which case, they are referred to as *silicon-based* PUFs. A ring oscillator is the simplest example of a PUF as it generates a distinct frequency for each chip it is implemented on.

PUFs are considered to be cryptographic primitives and can be used as a basic building block to construct security protocols and design secure systems. This primitive differs from those described in Sect. 1.4 in one crucial aspect, that is, the security of PUFs is based on the difficulty of replicating their exact physical structure rather than on the difficulty of solving a mathematical problem (e.g. factorising large integer number as the case of RSA asymmetric ciphers).

PUFs have a number of security-related applications, some of which are already being integrated into commercial products and the rest are still under development. We will briefly summarise below their current applications:

(a) Secure Key Storage

The majority of cryptographic algorithms (e.g. encryption, authentication, etc.) used in computing devices require a previously installed key, which act as the root of trust. This key is typically generated externally and injected into the device's memory.

This approach is vulnerable to security threats such as memory read-out and data remanence attacks. It also typically requires the manufacturers of the device to handle the key insertion process, which can be also a security risk.

PUF technology provides an alternative approach for key generation and storage, wherein the response of a PUF circuit is used to construct a root key. This approach removes the need for storing the key in an on-chip memory, which provides better protection against the above-mentioned attacks; it also removes the need to trust the manufacturers to handle the keys.

(b) Low-Cost Authentication

The relatively simple structure of PUF design makes it an attractive choice for low-cost authentication schemes, wherein the unique challenge/response behaviour of a PUF is used as an identity of physical objects. This is especially useful for recourse-constrained systems such as Internet of things devices which cannot afford classic security solutions [20].

(c) Anti-counterfeiting Design

PUF technology can be employed to limit overproduction of integrated circuits by malicious factories, which is causing significant financial losses to design houses every year.

This is achieved by embedding each chip with a PUF circuit that locks the design. At the post-fabrication stage, the behaviour of each PUF is characterised and authenticated by the design house; the latter can generate a passkey to activate only authenticated chips [27].

(d) Secure Hardware Tokens

PUFs can also be used in hardware-assisted cryptographic protocols, thanks to their complex challenge/response behaviours, which make them intrinsically more tamper-resistant than other tokens that rely on digitally stored information (e.g. smart cards and secure memory devices) [28].

(e) Secure Remote Sensing

PUF circuits are susceptible to environment parameter variations such as temperature or supply voltage, which effectively means that the response of a PUF depends on both the applied challenges and on the ambient parameters; this makes it feasible to use the PUF as a sensor to measure the change in the environment conditions. To do this, the correlation between the PUF responses and the physical quantities being measured should be characterised before the sensor is deployed. This approach makes it feasible to construct low-cost secure remote sensing schemes as it removes the need for implementing a separate encryption block on the device, because the readings generated by PUF cannot be understood by an adversary who has not characterised the behaviour PUF circuits; there are several examples of PUF-based secure sensing schemes proposed in the literature [29–31].

Currently, there are a number of companies driving the development of PUF-based security solution in the above application areas, and they are also exploring other usages [32–34].

1.7   Conclusions

We have provided a summary of the existing security attacks; we have also explained briefly the cryptographic primitives used to defend against such threats; in addition, we have explained that hardware security is becoming a major challenge; this is driven by a number of factors including the global nature of integrated circuit design chain, which brings about increased risks of counterfeiting and hardware Trojans. In addition, there are an increasing number of applications which relies on the use of trusted hardware platforms such as Internet of things, multiparty computation and secure online banking. Finally, we have introduced the concept of physically unclonable functions and explained how this technology can help to overcome some of the challenges of designing secure and trusted computing systems.

1.8   Problems

1. What are the main security attacks on electronics systems?
2. What are the main factors a designer needs to consider when developing security defence mechanisms?
3. Why is it challenging to validate the security of an electronic system such that exists in smart mobile phones?
4. What is the main requirement a hash function needs to satisfy in order to be suitable for security-sensitive applications?
5. What is a hardware Trojan?
6. Name three malicious functionalities a hardware Trojan may perform.
7. What is a physically unclonable function?
8. Explain the difference between physically unclonable functions and other cryptography primitives such as symmetric ciphers.

9. Can physically unclonable functions be used to generate a nonce? Explain your answer.
10. Explain how physically unclonable functions can be used as a secure sensor.

References

1. A. Zanella, N. Bui, A. Castellani, L. Vangelista, M. Zorzi, Internet of things for smart cities. IEEE Internet Things J. **1**, 22–32 (2014)
2. K. Nohl, J. Lell, BadUSB: on accessories that turn evil, Security Research Labs. Black Hat USA Presentation (2014)
3. R. Poroshyn, *Stuxnet: The True Story of Hunt and Evolution* (Createspace Independent Pub, 2014)
4. M. Tehranipoor, F. Koushanfar, A survey of hardware Trojan taxonomy and detection. IEEE Des. Test Comput. **27**, 10–25 (2010)
5. M. Rostami, F. Koushanfar, R. Karri, A primer on hardware security: models, methods, and metrics. Proc. IEEE **102**, 1283–1295 (2014)
6. B. Halak, J. Murphy, A. Yakovlev, Power balanced circuits for leakage-power-attacks resilient design. Sci. Inf. Conf. (SAI) **2015**, 1178–1183 (2015)
7. C. Clavier, J.S. Coron, N. Dabbous, Differential power analysis in the presence of hardware countermeasures, in *Proceedings of the Second International Workshop on Cryptographic Hardware and Embedded Systems,* vol. 1965 LNCS (2000), pp. 252–263
8. M.L. Akkar, Power analysis, what is now possible, in *ASIACRYPT* (2000)
9. S. Skorobogatov, Data remanence in flash memory devices, in *Presented at the Proceedings of the 7th International Conference on Cryptographic Hardware and Embedded Systems* (Edinburgh, UK, 2005)
10. S.P. Skorobogatov, R.J. Anderson, Optical fault induction attacks, in *Cryptographic Hardware and Embedded Systems—CHES 2002: 4th International Workshop Redwood Shores, CA, USA, August 13–15, 2002 Revised Papers*, ed. by B.S. Kaliski, ç.K. Koç, C. Paar (Springer Berlin Heidelberg, Berlin, Heidelberg, 2003), pp. 2–12
11. E.F. Foundation, *Cracking DES: Secrets of Encryption Research, Wiretap Politics & Chip Design* (Electronic Frontier Foundation, 1998)
12. J. Daemen, V. Rijmen, *The Design of Rijndael: AES—The Advanced Encryption Standard* (Springer Berlin Heidelberg, 2013)
13. D.R. Stinson, Universal hashing and authentication codes, in *Advances in Cryptology—CRYPTO '91: Proceedings*, ed. by J. Feigenbaum (Springer Berlin Heidelberg, Berlin, Heidelberg, 1992), pp. 74–85
14. S. Even, O. Goldreich, A. Lempel, A randomized protocol for signing contracts. Commun. ACM **28**, 637–647 (1985)
15. C.-K. Chu, W.-G. Tzeng, Efficient k-Out-of-n oblivious transfer schemes with adaptive and non-adaptive queries, in *Public Key Cryptography—PKC 2005: 8th International Workshop on Theory and Practice in Public Key Cryptography, Les Diablerets, Switzerland, January 23–26, 2005. Proceedings*, ed. by S. Vaudenay (Springer Berlin Heidelberg, Berlin, Heidelberg, 2005), pp. 172–183
16. M. Backes, A. Kate, A. Patra, Computational verifiable secret sharing revisited, in *Advances in Cryptology—ASIACRYPT 2011: 17th International Conference on the Theory and Application of Cryptology and Information Security, Seoul, South Korea, December 4–8, 2011. Proceedings*, ed. by D.H. Lee, X. Wang (Springer Berlin Heidelberg, Berlin, Heidelberg, 2011), pp. 590–609
17. T. Eccles, B. Halak, A secure and private billing protocol for smart metering, in *IACR Cryptology ePrint Archive,* vol. 2017 (2017), p. 654

18. S. Adee, The hunt for the kill switch. IEEE Spectr. **45**, 34–39 (2008)
19. S. Mitra. (2015, January 2) Stopping hardware Trojans in their tracks. *IEEE Spectr.*
20. W. Trappe, R. Howard, R.S. Moore, Low-energy security: limits and opportunities in the internet of things. IEEE Secur. Priv. **13**, 14–21 (2015)
21. C. Hazay, Y. Lindell, Constructions of truly practical secure protocols using standard smartcards, in *Presented at the Proceedings of the 15th ACM Conference on Computer and Communications Security* (Alexandria, Virginia, USA, 2008)
22. R. Canetti, Universally composable security: a new paradigm for cryptographic protocols, in *Presented at the Proceedings of the 42nd IEEE Symposium on Foundations of Computer Science* (2001)
23. B. Halak, S. Shedabale, H. Ramakrishnan, A. Yakovlev, G. Russell, The impact of variability on the reliability of long on-chip interconnect in the presence of crosstalk, in *International Workshop on System-Level Interconnect Prediction* (2008), pp. 65–72
24. D.J. Frank, R. Puri, D. Toma, Design and CAD challenges in 45 nm CMOS and beyond, in *IEEE/ACM International Conference on Computer-Aided Design* (2006), pp. 329–333
25. C. Alexander, G. Roy, A. Asenov, Random-dopant-induced drain current variation in nano-MOSFETs: a three-dimensional self-consistent Monte Carlo simulation study using (Ab initio) ionized impurity scattering. Electron Devices, IEEE Trans. **55**, 3251–3258 (2008)
26. L. Daihyun, J.W. Lee, B. Gassend, G.E. Suh, Mv Dijk, S. Devadas, Extracting secret keys from integrated circuits. IEEE Trans. Very Large Scale Integr. VLSI Syst. **13**, 1200–1205 (2005)
27. A. Yousra, K. Farinaz, P. Miodrag, Remote activation of ICs for piracy prevention and digital right management. IEEE/ACM Int. Conf. Comput.-Aided Design **2007**, 674–677 (2007)
28. U. Rührmair, Oblivious transfer based on physical unclonable functions, in *Trust and Trustworthy Computing: Third International Conference, TRUST 2010, Berlin, Germany, June 21–23, 2010. Proceedings*, ed. by A. Acquisti, S.W. Smith, A.-R. Sadeghi (Springer Berlin Heidelberg, Berlin, Heidelberg, 2010), pp. 430–440
29. Y.G.H. Ma, O. Kavehei, D.C. Ranasinghe, A PUF sensor: securing physical measurements, in *IEEE International Conference on Pervasive Computing and Communications Workshops (PerCom Workshops)* (Kona, HI, 2017), pp. 648–653
30. K. Rosenfeld, E. Gavas, R. Karri, Sensor physical unclonable functions, in *IEEE International Symposium on Hardware-Oriented Security and Trust (HOST)* (Anaheim, CA, 2010), pp. 112–117
31. H.M.Y. Gao, D. Abbott, S.F. Al-Sarawi, PUF sensor: exploiting PUF unreliability for secure wireless sensing. IEEE Trans. Circuits Syst. I Regul. Pap. **64**, 2532–2543 (2017)
32. Intrinsic-Id. (2017). Available: http://www.intrinsicid.com/products/
33. Verayo. (2017). Available: http://verayo.com/tech.php
34. Coherentlogix. (2017). Available: https://www.coherentlogix.com/products/hyperx-processors/security/

Physically Unclonable Functions: Design Principles and Evaluation Metrics

2

2.1 Introduction

Physically Unclonable Functions (PUFs) have recently emerged as an attractive technology for designing affordable secure embedded systems. A PUF can be seen as a physical entity, which has a relatively simple structure, and has a number of interesting characteristics relevant to security applications, namely; it is relatively easy to evaluate but practically impossible to predict, in addition, it is easy to make a PUF device but very hard to duplicate. The term PUF has been coined in [1] where authors also introduced the term silicon PUFs, which refers to physically unclonable devices built using conventional integrated circuit design techniques. Silicon PUFs exploit the inherent variations in manufacturing process among integrated circuits (ICs) with identical masks to uniquely characterize each IC.

It should be noted here that PUFs could, in principle, be constructed from any physical entity. Examples of non-silicon designs include optical PUFs that exploit the random scattering of light [2, 3], and radio-frequency (RF) designs that are based on the unique characteristic of the electromagnetic waves emitted from a device or an integrated circuit while operating [4–6]. There are more exotic constructions of secure object identification schemes such as those based on exploiting the microscopic imperfections in the surface of a piece of paper to generate a unique signature [7].

In this book, we will focus on silicon-based PUFs as it is the most likely candidate for building secure embedded systems at affordable costs. This chapter aims to:

1. Introduce the concept of physical disorder.
2. Explain the origin of physical disorder in integrated circuits.

© Springer International Publishing AG, part of Springer Nature 2018
B. Halak, *Physically Unclonable Functions*,
https://doi.org/10.1007/978-3-319-76804-5_2

3. Introduce different architectures of silicon based PUF and show how these can be constructed.
4. Outline the metrics used to evaluate a PUF designs.
5. Give the reader an insight into the design and implementation of PUF on configurable hardware platforms.

It is hoped that this chapter will give the reader the necessary skills and background to be able to construct their own PUF devices and evaluate their metrics.

2.2 Chapter Overview

The organisation of this chapter is as follows, Sect. 2.3 outlines the concept of physical disorder, in Sect. 2.4, we look more closely into the conditions under which integrated circuits can exhibit forms of physical disorder. Section 2.5 presents a generic framework to design a PUF device using integrated circuit design techniques. Examples of existing designs are discussed in a great depth in Sects. 2.6, 2.7 and 2.8 respectively. Section 2.9 summarises the important metrics employed to assess the quality and usability of PUF circuit architectures. Section 2.10 discusses in details the design and implementation processes of a configurable PUF architecture using filed programmable logic arrays (FPGAs). A comprehensive comparison of the characteristics of publically available ASCI implementation of PUFs is presented in Sect. 2.11. Learned lessons are summarised in Sect. 2.12. Finally, problems and exercises are included in Sect. 2.13.

2.3 What Is Physical Disorder?

Physical disorder refers to the random imperfections present in the structure of physical objects; these are typically observed at a microscopic scale. Examples of such fascinating randomness are many; they are present in both biological and physical entities around us. Take, for example, the rose petal shown in Fig. 2.1a; a close-up look at a piece of paper in Fig. 2.1b also reveals notable three-dimensional irregularities of its interleaved fibres. Another example is a microscopic image of a coffee bean in Fig. 2.1c that shows its uneven surface with three dimensional random structures, actually, if we were to look into our mouths

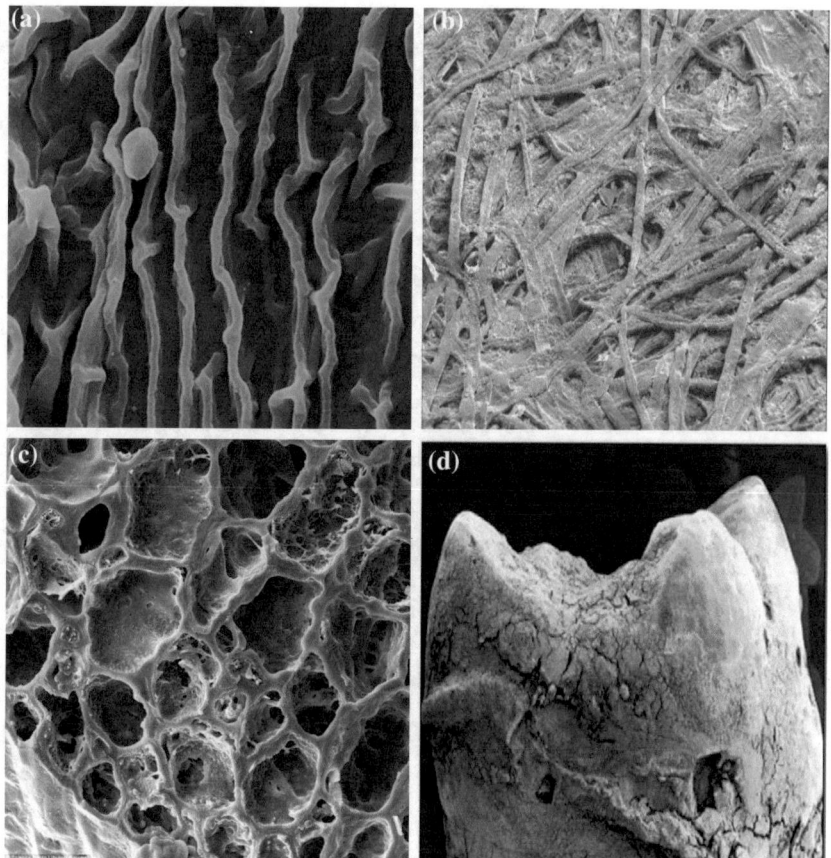

Fig. 2.1 Examples of physical disorder: **a** rose petal, **b** paper **c** a coffee bean **d** a tooth

we would find such disorder in abundance, Fig. 2.1d shows a microscopic image of a tooth, it may not be pretty, but it is certainly irregular.

More importantly, physical disorder can be found in abundance in modern integrated circuits, this is because the continuous scaling of the semiconductor technologies has made it extremely difficult to fabricate precisely sized devises, for example, Fig. 2.2 shows the irregular structure of the metal conductors in a semiconductor chip fabricated in 90 nm technology.

This physical disorder is unique for each device, it is also hard to replicate, therefore it can be used to give each physical object an identity.

Fig. 2.2 Physical disorder of integrated circuits

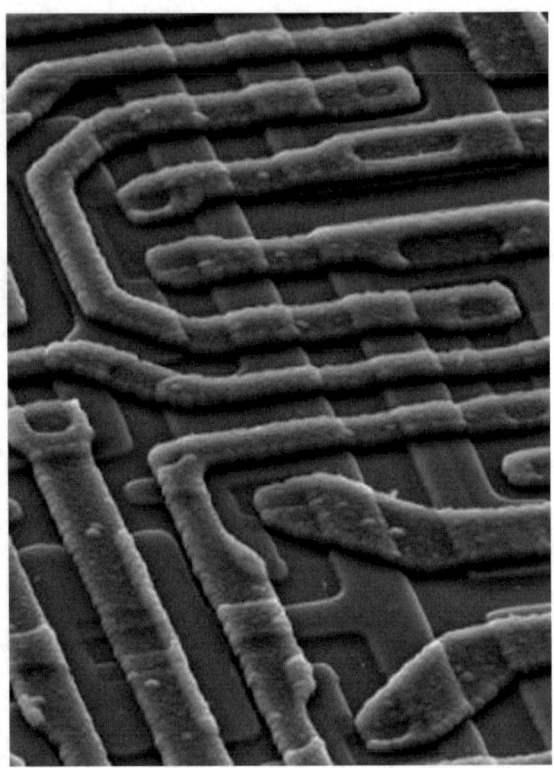

2.4 The Origins Physical Disorder in Integrated Circuits

In principles, integrated circuits are synthetic; therefore, it should be possible to design out all irregularities in their shapes or structures, however; this is not the case in the vast majority of modern chips, and the reason for this is variability. The latter refers to the inaccuracies in manufacturing processes and within-die voltage-temperature variations that lead to fluctuations in circuit performance and power consumption [8]. It arises from scaling very large-scale integrated (VLSI) circuit technologies beyond the ability to control specific performance-dependent and power-dependent parameters. Two sources of variations can be identified [9, 10], environmental factors, which include variations in power, supply voltage, operation temperature and degradation in the electrical parameters of devices also known as "aging". Second, physical factors which include variations in the dimensions and the structures of the fabricated devices. Although environmental factors may lead to fluctuations in the electrical parameters of an integrated circuit, they do not cause physical disorder; therefore, they will not be discussed further in this chapter. Physical sources of variability are mainly due to the fact that the achievement of parameter precision becomes exponentially more difficult, as technology scales down, due to the limitations imposed by quantum mechanics [8, 9], these factors can be attributed to different sources, namely: [8, 10–14].

2.4.1 Devices' Geometry

The first set of process variations relate to the physical geometric structure of MOSFET and other devices (resistors, capacitors) in the circuit. These typically include:

(a) Film thickness variations: The gate oxide thickness (T_{ox}) is a critical but usually relatively well controlled parameter. Variation tends to occur primarily from one wafer to another with good across wafer and across die control.

(b) Lateral dimension variations: Lateral dimensions (channel length, channel width) typically arise due to photolithography proximity effects or plasma etch dependencies. MOSFET are well known to be particularly sensitive to effective channel length (L_{eff}), as well as gate oxide thickness and to some degree the channel width. Of these, channel length variation often is singled out for particular attention, due to the direct impact such variation can have on device output current characteristics [11].

2.4.2 Devices' Material

Another class of process variations in MOSFETS relates to internal material parameters, including:

(a) Doping variations are due to dose, energy, angle, or other ion implant dependencies. Depending on the gate technology used, these deviations can lead to some loss in the matching of NMOS versus PMOS devices even in the case where within wafer and within die variations are very small.

(b) Deposition and Anneal: Additional material parameters deviations are observed in silicide formation, and in the grain structure of poly or metal lines. These variations may depend on the deposition and anneal processes. These material parameters' deviations can contribute to appreciable contact and line resistance variation.

2.4.3   Interconnects' Geometry

(a) Line width and line space (w, s): Deviations in the width of patterned lines again arise primarily due to photolithography and etch dependencies. Line width variations can directly impact line resistance, as well as the inter-layer capacitances; they can also result in differences in wire spacing affecting the magnitude of crosstalk glitches.

(b) Metal thickness (t): in conventional metal interconnect, the thickness of sputtered or otherwise deposited metal films and liners or barriers is usually well controlled, but can vary from wafer to wafer and across wafer. In damascene (e.g. copper polishing) processes, on the other hand, dishing and erosion can significantly impact the final thickness of patterned lines, which may lead to a large variation of metal thickness on the same wafer.

(c) Dielectric Height (h): The thickness of deposited and polished oxide films can also suffer substantial deviations. While wafer level deposition thickness can vary (typically on the order of 5%), more troublesome are pattern-dependent aspects of such deposition. Chemical mechanical planarization (CMP) process can also introduce strong variations across the die, resulting from the effective density of raised topography in different regions across the chip.

2.4.4   Interconnects Material

The variability of material properties can also be important in interconnect structures. In particular, we consider the following sources of variations:

(a) Metal resistivity (ρ): While metal resistivity variation can occur (and include a small random element), resistivity usually varies appreciably on a wafer to wafer basis and is usually well controlled.

(b) Dielectric constant (ε) may vary depending on the deposition process, but is usually well controlled.

(c) Contact and via resistance: Contact and via resistance, can be sensitive to etch and clean processes, with substantial wafer-to-wafer and random components.

The impact of variability is expected to be significant in future technologies [12], making variations an unavoidable characteristic of VLSI circuits. This means it is going to be harder to predict with certainty the exact performance and/or power consumption a particular implementation of a circuit is going to have, because variations in the physical dominations of devices will lead to variation in the their electrical parameters. Figure 2.3 shows the magnitude of variation in device

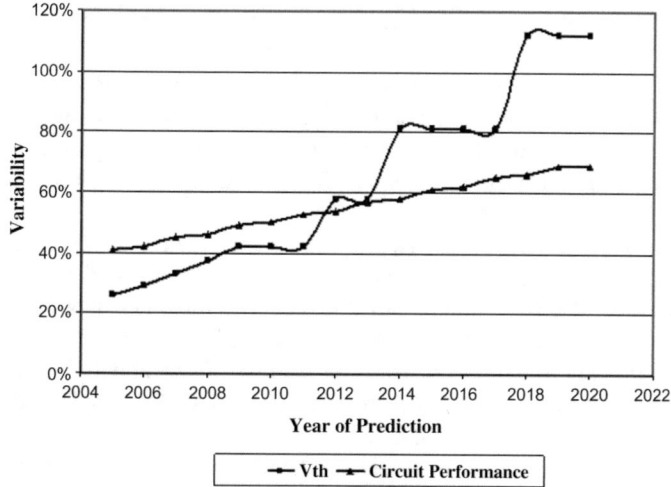

Fig. 2.3 The impact of variability on the electrical parameters of VLSI circuits

threshold voltage (V_{th}) and circuits' performance (measured as a function of wires and devices delays). The data are obtained from ITRS report on design [12].

Such uncertainty generally means a designer needs to consider the worst-case scenario, which may lead to sub-optimal designs, take for example a case where a thousand processor cores are designed to run at a maximum frequency of 600 MHz, and assuming the manufacturing process leads to 10% variability in performance. If we assume, for simplicity, that the performance (measured as the maximum operating frequency) has a Gaussian distribution with a mean of 600 and a standard deviation of 20. This means that only half of the fabricated processors are able to run at the maximum intended speed (600 MHz).

In fact, the operation frequency will have to be reduced to 540 MHz in order to ensure correct functionality of ALL fabricated devices and avoid potential timing errors.

Such wort-case scenario approach is costly but unavoidable in the vast majority of cases.

However, it is not all bad news, as we will see in the next section, process variations can be exploited to design a physically unclonable function, a promising cryptographic primitive that may hold the answer to many of the security challenges facing resource-constrained embedded systems.

2.5   How to Design a Physically Unclonable Function

A Physically Unclonable Function can be regarded as a functional entity embedded into a physical object such as a silicon chip. The output (*aka* the response (*R*)) of this entity is dependent on its input (*aka* the challenge (*C*)) and the intrinsic physical properties of the host system. Such entity should be hard to clone even if its detailed physical structure is revealed.

One example of such a function is the conceptual structure in Fig. 2.4; it is based on a basic ring oscillator circuit. This structure can be thought of as a PUF device that can be activated by setting the enable signal to logic (1), as a result, it starts oscillating at a frequency that depends on both the structure of the circuit (e.g. number of inverters) and the physical properties of the underlying technology, which determines the speed of these inverters. Such a PUF has only one challenge (Enable is high) and one response (the oscillation frequency). The presence of physical disorder in modern semiconductor devices, as we have seen earlier, means that this PUF can generate different frequency responses when implemented on multiple devices, these responses can be used as a hardware signature unique for each physical implementation of this function.

The unclonability of this conceptual example is achieved, thanks to the inherent physical limitations of current chip fabrication processes as discussed in the previous section.

Although the PUF technology is relatively new, the idea of identifying objects based on their unique set of physical properties is not, in fact, the ancient Babylonian used fingerprints to protect against forgery. At that time legal contracts which were written on a piece of clay needed to be signed by contracting parties to confirm their authenticity, they did this by impressing their fingerprints onto the same piece of clay [15]. More recently, the "uniqueness" of physical objects has been used to identify nuclear weapons during the cold war; to do this a thin layer of coating is sprayed onto the surface of the device to give it unique light-reflecting characteristics. To identify a nuclear arm, its surface would be illuminated again with the same light and the resulting interference patterns is compared with previously stored patterns [16].

These examples indicate that in order to design a physically unclonable function, one needs two ingredients; the first is an inherent physical disorder unique to each object, and the second is a method to measure such physical disorder, quantify it and store it.

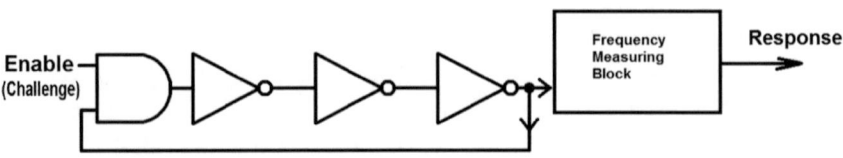

Fig. 2.4 A conceptual example of a PUF device

In the case of Babylonian clay fingerprints, physical disorder is measured by visual inspection, whereas coating-based identification of nuclear arms requires more sophisticated equipment to analyze light interference patterns.

Modern intergraded circuits have an abundance of physical order, thanks to process variations, so the first ingredient is easy to obtain, however; it is more complicated to transform such disorder to a measurable quantity. There are three basic parameters, which are typically measured in integrated circuits, namely: current (I), voltage (V) and delay (D). Other metrics are derivations of these three measures, for example; resistance of a metal conductor is estimated by dividing the voltage drop across its two ends over the average current passing through it.

Based on the above discussion, the first requirement of a silicon PUF device is formulated as follows:

> *Given a semiconductor technology with a known degree of process variation (var), the PUF circuits need to transform such variations into a measurable quantity (mq) (e.g. voltage, delay or current).*

Another expectation of the PUF circuit is to provide a distinctive response to each applied challenge; such response in the vast majority of application needs to have a digitized form, therefore the second requirement of a PUF circuit is formulated as follows:

> *The PUF circuit needs to transform its measurable quantity (mq) into a digitized form (i.e. a binary response).*

Given the above two requirements, one can develop a generic architecture for silicon PUF circuits as shown in Fig. 2.5.

By comparing Figs. 2.4 and 2.5, we can deduce that the ring oscillator is a transformation block and the frequency-measuring block is a conversion block. The former turns the challenge (the enable signal) and the process variation of the implementation technology (*var*) into a measurable quantity (i.e. the oscillation frequency that is the inverse of delay), and the latter turns the measured frequency into a binary value (i.e. the response).

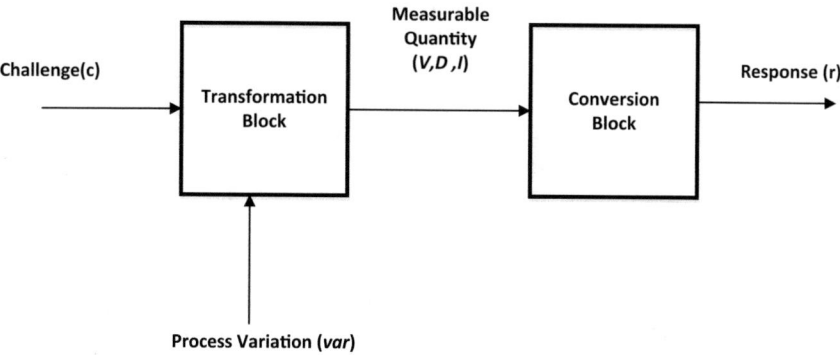

Fig. 2.5 A generic architecture for silicon-based PUFs

This generic representation of the PUF architecture makes it easier to explore the design space and re-use existing circuitry; this is because the design problem of the PUF can now be thought of as designing two separate modules: a transformation block and a conversion circuit, examples of both of which are widely available in the literature. Examples of transformation blocks include digital to analogue converters, ring oscillators, and current sources. Conversion circuits may include analogue to digital converters, time to digital converters, phase decoding circuitry [17] and comparators [18, 19].

The generic architecture presented in Fig. 2.5, gives us an intuitive way to classify existing implementation of silicon-based PUF devices based on the measurable quantity as follows, voltage-based, current-based and delay-based PUFs.

It should be noted that that there exist PUF constructions which do not fall entirely under any of these three categories, such as those which exploit variations in both driving current and threshold voltages. Nevertheless, this categorization provides an instinctive method to approach the design problem of PUFs.

In the next section, we will review in details existing PUF circuits and discuss their different implementations.

2.6 Delay-Based PUFs

These structures transform process variations into a measurable delay figure, and the latter is converted into a binary response. Initial designs of PUF circuits have all been based on this structure, there are many examples of such constructions in the literature, including arbiter-based circuits [1, 20–22], ring oscillator designs [23–31] and those based on asynchronous structure [31]. We will discuss examples of these constructions in the following subsections.

2.6.1 Arbiter PUFs

Let us consider the arbiter-based structure shown in Fig. 2.6; it consists of two digital paths with identical nominal delay and an arbiter. When an input signal is

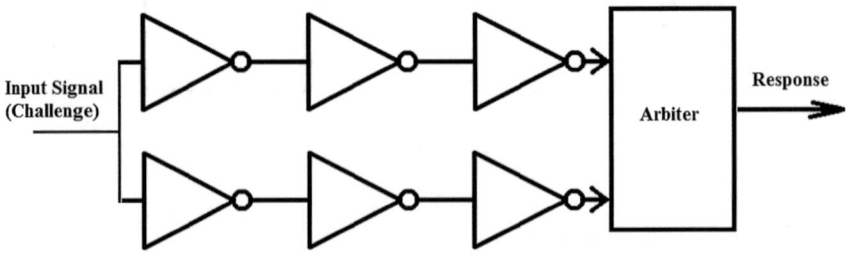

Fig. 2.6 A single challenge arbiter PUF

Fig. 2.7 Circuit diagram of an arbiter based on an S-R latch

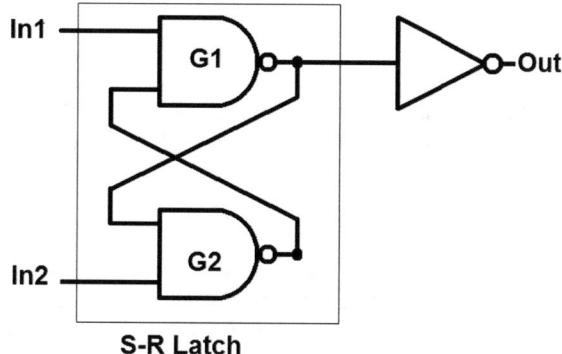

S-R Latch

applied, it will propagate through these two paths, and arrive to the inputs of the arbiter at slightly different moments because the intra-die variability, the arbiter outputs a logic "1" or a logic "0" depending on which path wins the race.

When such structure is implemented on various chips, the response will vary because of inter-die variations, in other words, it is unique to each implementation, therefore, it can be used as a hardware signature.

The arbiter circuit is normally designed using a Set-Reset latch; where in the latch is constructed using two cross-coupled gates as shown in Fig. 2.7. The operation principles of this circuit are as follows; when both inputs are low, the output will also be low, if only one input goes high, the output will change accordingly and become locked such that all changes on the other input do not affect the status of the output. In this example, if the signal on inputs (*In 1*) goes high, then the output will go high and remain high, if the signal on input (*In 2*) goes high, then the output will remain low. If the signals on both inputs go high within a short period of time of each other's, the output will assume a value based on the signal that arrives first (i.e. "1" if (*In 1*) is first to arrive, "0" if (*In 2*) is first to arrive). However; if both signals arrive within a very short period, the output may enter a metastable state for an indefinite amount of time. More precisely until the output of the top gate (G1) crosses the threshold voltage of the bottom gate (G2), in which case the arbiter output becomes high, it is also possible that the output of bottom gate (G2) crosses the threshold voltage of the top gate first, in which case the output of the arbiter remains low indefinitely.

Assuming metastability happens at (t = 0), the output voltage of (G1) at time (*t*) will be given as [32]:

$$V(t) = V_0 e^{\frac{t}{\tau}} \tag{2.1}$$

where:

τ is a technology parameter that depends only on the circuits characteristics.

Metastability ends approximately when the output voltage of the two gates (G1 or G2) reaches the threshold voltage of the NAND gate, by substituting V with V_{th} in the previous equation we get:

$$t = \tau \ln \frac{V_{th}}{V_0} \qquad (2.2)$$

Probabilistic analysis shows that given a metastability event at (t = 0), the probability of metastability at $t > 0$ (P_m) can be calculated as

$$P_m(t) = e^{-\frac{t}{\tau}} \qquad (2.3)$$

So in theory, it is possible to calculate exactly when the metastability event is going to end but that requires detailed knowledge of the circuit geometries and the voltage level at the output of the two gates (G1 and G2) when the metastability event starts. Such information is very hard to obtain in practice. The problem of the above scenario is that it will be very difficult to work out whether the output of the arbiter is low because the signal (*In 2*) has arrived first or because the cross-coupled gates are still in the metastable state.

One of the earliest examples to appear in the literature of an arbiter-PUF is by Daihyun et al. in [1], in which the authors proposed a multi-bit challenge design that consists of a number configurable multiplexer pipelined in a serial manner. Each multiplexer has a select input, which can select one of possible digital paths to connect its primary inputs to its outputs.

The number of bits of each challenge corresponds to the number of connected multiplexers; each challenge will lead to unique configuration of the digital paths, hence a unique response. An example of a two bits arbiter based on such a design is shown in Fig. 2.8.

The maximum number of challenge/response pairs (*CRP*) which can be obtained using an arbiter PUF that has k stages, each of which is controlled by one selection bit, is given by the following equation:

$$CRP = 2^k \qquad (2.4)$$

The structure proposed of the arbiter in the above design was based on a transparent latch, however; as reported by the authors, the asymmetric nature of such a latch can greatly affect the predictability of the response, more specifically,

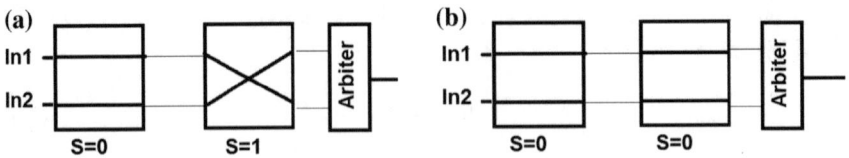

Fig. 2.8 Structure of a two bit challenge arbiter PUF: **a** the challenge is "01" **b** the challenge is "00"

the arbiter tends to always favour one path over the other, which meant that 90% of the responses are "0" [1]. Proposed techniques to tackle this problem, by systematically changing the delay of logic paths, may reduce the effects of this problem but will make the behaviour of the arbiter PUF inherently more predictable, because they are not based on the physical disorder but on designed variations.

The difficulty in achieving a perfectly symmetric design is greatly increased by the fact that layouts of the vast majority of modern chips are produced using automated place and route software tools (e.g., designs based on standard cell logic libraries or those placed on configurable hardware platforms). In these cases, the degree of freedom a designer has to control the layout is reduced.

Therefore, manual layout tools need to be used for designing arbiter PUFs to ensure the symmetry of their delay paths.

2.6.2 Ring Oscillator PUF

The original ring oscillator based PUF is made of two multiplexers, two counters, one comparator and K ring oscillators [33]. Each ring oscillates at a unique frequency depending on the characteristics of each of its inverters, the two multiplexers select two ROs to compare. The two counter blocks count the number of oscillations of each of the two ROs in a fixed time interval. At the end of the interval, the outputs of the two counters are compared, and depending on which of the two counters has the highest value, the output of the PUF is set to 0 or 1. A block diagram of its structure is shown in Fig. 2.9 [33].

Similarly to the arbiter PUF, this design still requires the ring oscillator to have identical nominal delays, however it removes the need for an arbitration circuit, therefore overcoming the problem of metastability, which allows for higher reliability.

The maximum number of challenge/response pairs (*CRP*) obtained from an RO PUF with (k) ring oscillators, is given as below:

$$CRP = \frac{k.(k-1)}{2} \tag{2.5}$$

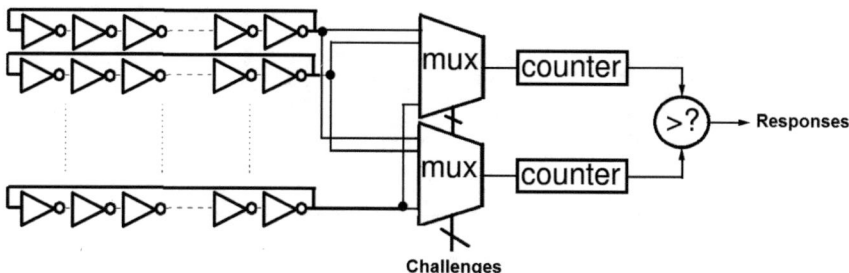

Fig. 2.9 Generic structure of ring oscillator PUFs

In comparison with arbiter-based designs, RO PUFs are larger and consume more power but they are less susceptible to metastability-induced errors. Both of these designs are vulnerable to machine learning attacks as will be seen later in Chap. 5.

2.6.3   Self-timed Rings PUF

This is another variation of delay based PUFs that has been proposed recently by Murphy et al. in [34], it has the same structure as the generic ring oscillator shown in Fig. 2.9, however it uses self-timed rings instead of the inverter chains. The design is based on the use of the Muller's C-element, a fundamental building block of asynchronous circuits. Figure 2.10 shows a CMOS implementation of the C-element based on the use of weakly inverted output, its operation principles are simple, basically the output will be set to a logic "1" or "0" when both inputs assume the logic "1" or logic "0" values respectively, otherwise the output retains its previous value (i.e. remains unchanged).

An example of a three stage self-timed ring (STR) is shown in Fig. 2.11. Each stage consists of a C-element and an inverter connected to one input of the C-element, which is marked R (Reverse), the other input is connected to the previous stage and marked F (Forward).

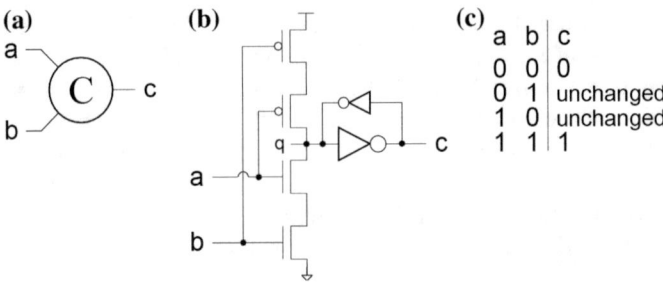

Fig. 2.10 The muller's C-element: **a** graphic representation **b** CMOS implementation **c** truth table

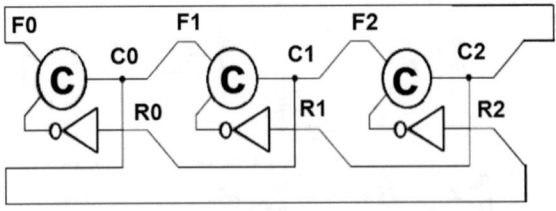

Fig. 2.11 A three stage self timed ring (STR)

To aid the understanding of the behaviour of this structure, we will introduce the concept of *Bubbles and Tokens*. A stage in the STR is said to contain a *"Bubble"* if its output is equal to the output of the previous stage. A stage in the STR is said to contain a *"Token"* if its output is different from the output of the previous stage.

Unlike inverter chains, self-timed rings do not oscillate unless the following conditions are satisfied. First, the number of stages (*STR*) should be at least three and second it should be equal to the sum bubbles and tokens contained in all the stages. These conditions are summarized in the equations below.

$$STR \geq 3 \tag{2.6}$$

$$STR = N_B + N_T \tag{2.7}$$

where

$N_B \geq 1$ is the number of bubbles

N_T is a positive even number of tokens

In order to meet such conditions the inputs of the self time rings should be initialised to the correct values otherwise it will not oscillate (i.e. it remains in deadlock state).

The main advantage of self-timed rings is that they increase the robustness of the PUF's responses against environmental variations, this comes at the cost of an increase in the cost of silicon area, moreover, these self-timed structures are prone to entering deadlock states.

2.7 Current-Based PUFs

These structures are capable of transforming random process variations into a measurable current figure; the latter is then converted into a binary response, unique for each device. We are going to study two examples of this type, the first is based on the use of sub-threshold currents generated by a stand-alone transistors' array, and the second is based on capturing the subtle variations in the leakage current of a Dynamic Random Access Memory (DRAM) cell.

2.7.1 Current-Based PUFs Using Transistors Arrays

One of the earliest examples of this type of PUFs can be found in [35], where the authors presented a MOSFET based-architecture. The main motivation behind this design is to harness the exponential dependency of the current in the sub-threshold region on threshold voltage (V_{th}) and gate-to-source voltage (V_{GS}), in order to increase the unpredictability of the PUF's behavior. A conceptual architecture of this design is shown in Fig. 2.12.

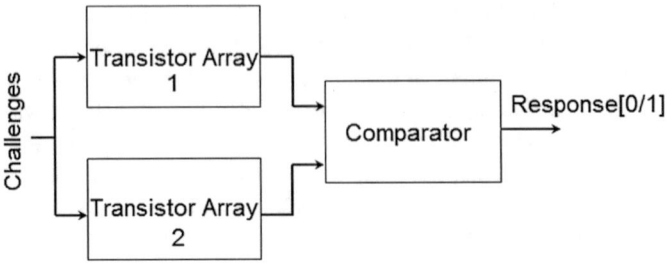

Fig. 2.12 A current-based PUF using transistor arrays

The operation principles of this design are as follows, the challenge bits are applied simultaneously to the inputs of two identically sized arrays of transistors, each challenge selects a number of transistors to be connected to the output of each array. The two outputs are then compared to generate a binary response. The number of challenge response pairs for this design is given below.

$$CRP = 2^{kn} \qquad (2.8)$$

where k, n are the number of columns and rows in the transistor arrays respectively.

There are a number of ways the transistors arrays can be designed, the proposed architecture in [35] suffers from a number of shortcomings, including the low level output voltage of the array, which makes it harder to design a suitable comparator. An improved version of the arrays 'design was proposed in [36]', a simplified block diagram of the same is shown in Fig. 2.13.

The array in Fig. 2.13 consists of k columns and n rows of a unit cell (as highlighted in red in Fig. 2.13). Each unit consists of two parts, and each part consists of two transistors connected in parallel, one of them has a minimally sized length in order to maximize the variability in its threshold voltage, we call it the "stochastic" transistor (e.g. N11x). The second transistor acts as a switch, either to remove the impact of stochastic transistor when it is ON, or to include when it is OFF (e.g. N11), we call it the "Switch transistor".

Each bit of the challenges is applied to a NMOS unit cell and its symmetric PMOS cell; take for example the challenge bit (c11), which is applied to the two units highlighted in red and green in Fig. 2.13.

If the challenge is logic "1", the "switch" transistors in the NMOS cell will be ON and those in the PMOS cell will be OFF. On the other hand, there is always a contribution from stochastic transistor regardless whether the challenge bit is '0' or '1', because the non-inverted and inverted versions of each input bit are connected to a stochastic transistor, therefore, regardless of the challenge bit, one out of two stochastic transistors will be selected and be part of the network. This architecture is, therefore known as "Two Chooses One" or "TCO".

The output of each array is dependent on the accumulated current flowing through the network of transistors.

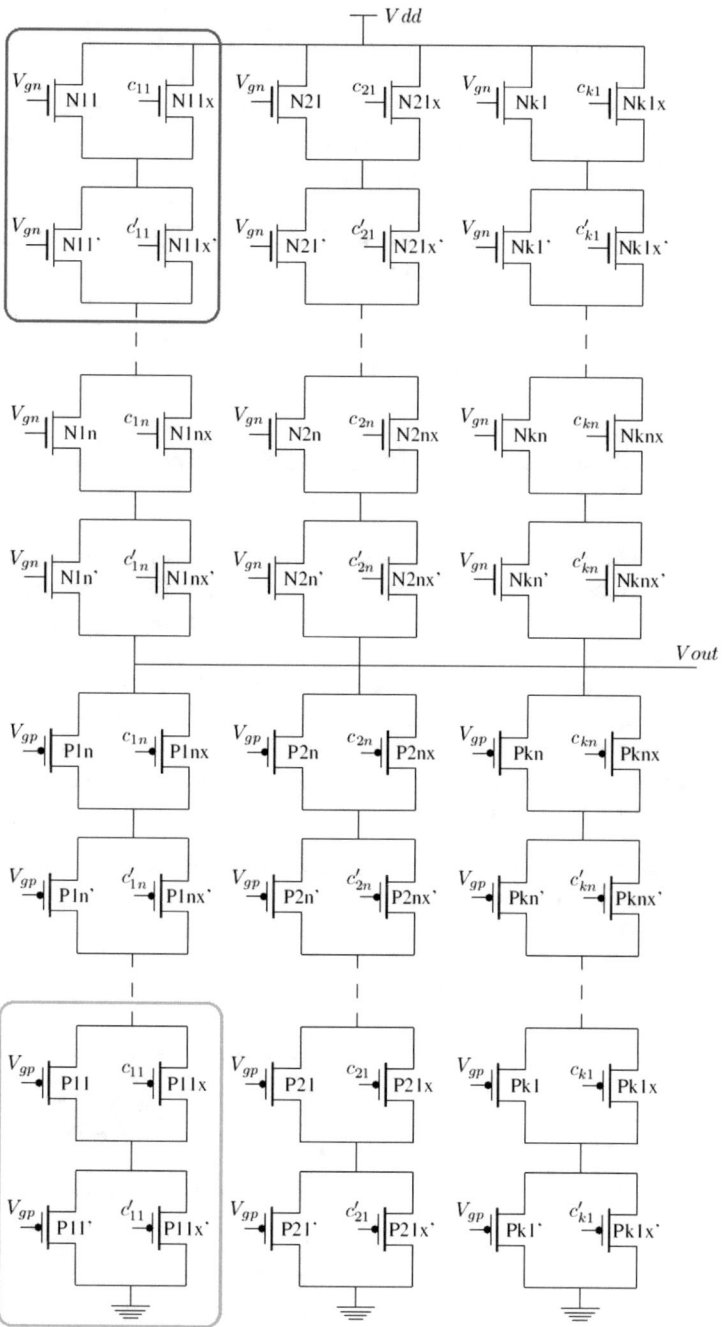

Fig. 2.13 An exemplar transistor array for a current-based PUF

The inherent intra-die variations ensure the voltage output of one of the array is slightly more than the other; a conventional dynamic comparators (e.g. Op-Amp-based comparators) can capture such a difference.

There are a number of considerations that need to be taken into account when building a current-based PUF as the one shown above. First, a proper biasing needs to be applied for *Vgn* and *Vgp* to ensure the stochastic transistors always operate in the sub-threshold region. Second, for the "Switch" transistors, their sub-threshold current should be negligible, and they should provide small ON-state resistance. As a rule of thumb, the dimensions (width and length) of the switch transistors should be 10 times of those of their corresponding stochastic transistors. Finally, the difference between the output voltages of the two arrays should be large enough for the comparator to detect it; otherwise, the behavior of the device will be unreliable.

2.7.2 Current-Based PUFs Using Dynamic Random Access Memories

A typical DRAM cell consists of a capacitor and an access transistor as shown in Fig. 2.14, such a cell is capable of storing a single data bit. The charge stored in the capacitor decays overtime due to the leakage current; therefore periodic refreshments are needed to preserve the contents of the memory. The rate at which the charge decays is proportional to the leakage current, which depends on the fabrication technology, and this varies from one cell to another and form one chip to another because of random process variations. A detailed process to exploit the unique decay behavior of DRAM cells to construct a PUF is presented in [37], a summary of this process is outlined below:

(1) A region of the memory is chosen to produce the PUF response, it is defined by the starting address and the size
(2) The refresh functionality is disabled for this reserved region
(3) An initial value is written into this region

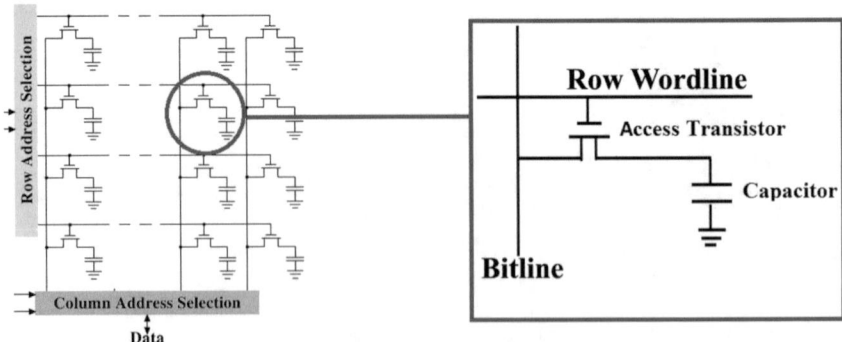

Fig. 2.14 A generic structure for a DRAM cell

(4)   Access is disabled to all cells in the reserved region for a chosen period of time
      (t), during which the charge in each cell decays at a unique rate proportional to
      its leakage current
(5)   Once the decay time has expired, the contents of the reserved region is read as
      normal to produce the response of the PUF
(6)   Normal operation is then resumed and the reserved region is made available
      again to the operating system.

The number of challenge/response pairs such a PUF can produce is dependent
on the size of the DRAM and the decay behavior of its underlying implementation.
For example, envisage a DRAM PUF with R possible reserved regions, where it is
possible to have N possible decay periods, each will lead to a unique response.

$$CRP = R \times N \qquad (2.9)$$

A number of important issues should be considered before adopting this
architecture; firstly, the decay behavior of the DRAM should be studied carefully to
allow optimal choice for the decay time. For example, if the implementation
technology has a very high leakage current then DRAM cells in the reserved region
may all lose their charges; as a result, the contents of the decayed cells may all
change. On the other hand, if the decay rate is very slow, then the initial contents of
the memory may not change at all. In both cases, the behavior of the resulting PUF
will be too predictable. Secondly, the minimum decay period should be defined
such that it allows the data contents of the reserved region to become distinctively
different from the originally stored values (i.e. at least the content of one memory
cell should change).

In comparison with the transistors-array-based design, a DRAM PUF may
require less area overhead as it can be built using pre-existing on-chip memory
blocks.

2.8   Voltage-Based PUFs

This type of design can transform process variations into a measurable voltage
figure; we are going to study two examples; the first is based on the use of static
random access memories (SRAM) and the second is based on the use of SR latches.

2.8.1   SRAM PUFs

PUFs based on Static Random Access Memories (SRAM) are one of the earliest
designs to appear in this category in [38], they have been initially proposed to
secure FPGA designs by generating a device-specific encryption key to scramble
the bit streams before storing it in an external memory. Such an approach protects

against a threat model where in an adversary is able to extract the bit stream by analysing the signal on the cable used to load the design, the use of SRAM PUF prevents such an adversary form re-using the same bit stream to program other FPGAs.

We are now going to discuss in more details the principles of an SRAM PUF. Figure 2.15 shows a typical SRAM cell, the latter is composed of two cross-coupled inverters (P1, N1, P2, N2) and two N-type access transistors (N3, N4). The two inverters have two stable states logic '1' or logic '0', which can be accessed through N3 and N4 by two bit lines namely 'bit' and 'bit_bar' (the compliment of 'bit'). Each inverter drives one of the two state nodes, Q or Q'. The access operation is in the control of one word line labeled as *WL*.

When this cell is powered up, the two cross-coupled inverters enter a "power struggle"; the winner will be ultimately decided by the difference in the driving strength of the MOSFETs in the cross-coupled inverters. Essentially, the SRAM cell has three possible states, two of these are "stable" and the third is "metastable" as shown in Fig. 2.16. If the transistors in the cross-coupled inverters circuits are perfectly matched, then the SRAM may remain in a metastable state "forever" when first powered up. In reality, although those transistors are designed to have identical nominal sizes, random variations in the silicon manufacturing process ensure that one inverter has a stronger driving current than that of the other inverter, this helps define the initial start-up value for the cell.

The majority of SRAM cells have preferred cell-specific initial state, which they consistently assume when powered up, this characteristic of SRAM memories allows them to be used for PUF constructions. The number of challenge/response pairs obtained from an SRAM-based PUF is proportional to the size of the memory,

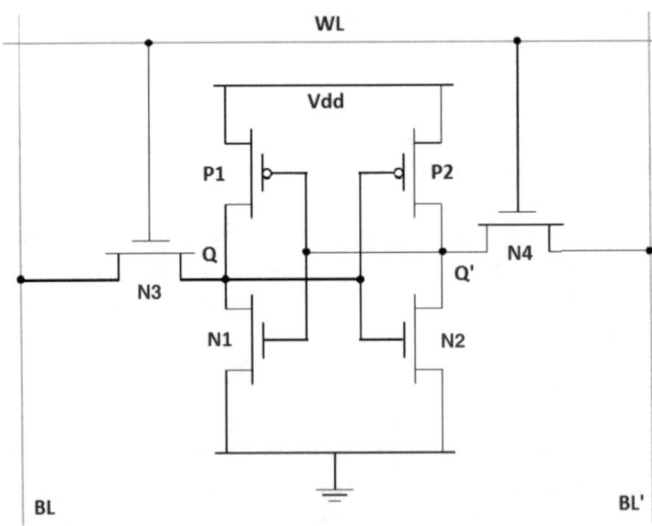

Fig. 2.15 A generic 6-transistor SRAM cell

Fig. 2.16 Inverter
characteristics for SRAM cell

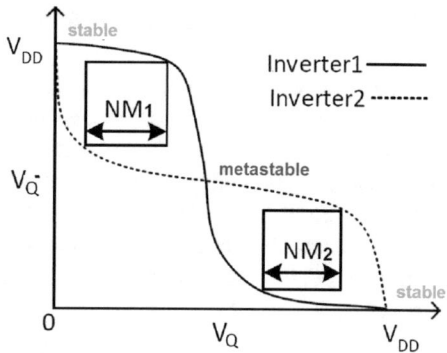

in this case, the challenge will be the reading address of the memory and the response is the start-up values of the addressed cells. For example, a 64 megabits Byte-addressable memory has 8 megabits of challenge/response pairs.

2.8.2 Latch-Based PUFs

This is another variation of voltage-based designs wherein the subtle variability of the threshold voltages of a cross-coupled NOR gates is exploited to generate a unique response [39, 40]. A basic design for this type is depicted in Fig. 2.17, it works as follows, let us assume an initial state (In = 0, Q = 1), when a logic "1" is applied to the input; the latch is forced to a metastable state, which can last for undetermined amount of time (please see Sect. 2.6.1 for more details on metastability). When the metastability finally resolves the output assumes a value of "0" or "1" depending on the variations of the driving strengths of the two gates.

Despite its simplicity, this design may not be very reliable as it intentionally invokes metastability events, the period of which is hard to predict, in addition the final output of the circuit may greatly be influenced by power supply variations and other environment conditions.

Fig. 2.17 A single bit
challenge/response latch PUF

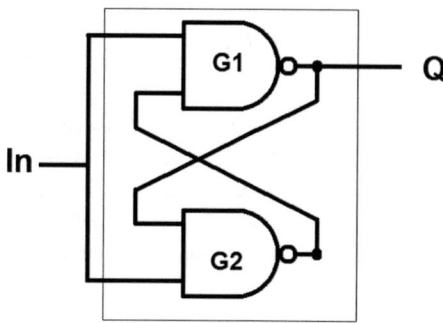

2.9   Evaluation Metrics of PUF Devices

In the previous sections, we discussed a number of possible designs for silicon-based physically unclonable functions; in this section, we introduce a number of metrics used to evaluate the quality of a PUF design, and more importantly its suitability for a specific application. We are going to introduce four metrics: uniqueness, reliability, uniformity, and tamper resistant.

The computation of these metrics relies on the concept of Hamming distance and Hamming Weight, therefore, it is appropriate to define these two terms before we proceed.

Definition 2.1

(a) **Hamming Distance** *The Hamming distance $d(a, b)$ between two words $a = (a_i)$ and $b = (b_i)$ of length n is defined to be the number of positions where they differ, that is, the number of (i)s such that $a_i \neq b_i$.*

(b) **Hamming Weight** *Let 0 denotes the zero vectors: $00 \ldots 0$, The Hamming Weight $HW(a)$ of a word $a = a_1$ is defined to be $d(a, 0)$, the number of symbols $a_i \;!= 0$ in a.*

2.9.1   Uniqueness

It is a measure of the ability of a device to generate unique identification, in other words. It is the measure of the ability of one PUF instance to have a uniquely distinguishable behavior compared with other PUFs with the same structure implemented on different chips.

The uniqueness metric is evaluated using the 'Inter-chip Hamming Distance' If two chips, i and j $(i \neq j)$, have n-bit responses, $R_i(n)$ and $R_j(n)$, respectively, for the challenge C, the average inter-chip HD among k chips is defined as [41]:

$$HD_{INTER} = \frac{2}{k(k-1)} \sum_{i=1}^{k-1} \sum_{j=i+1}^{k} \frac{HD\big(R_i(n), R_j(n)\big)}{n} \times 100\% \qquad (2.10)$$

Take for example the two instances of a PUF depicted in Fig. 2.18, which are assumed to be implemented on two different chips. When a challenge (011101) is applied on both circuits, each PUF outputs a distinct response. the Hamming distance between these is 2, this means 25% of the total response bits are diffident. Ideally Uniqueness should be close to 50%.

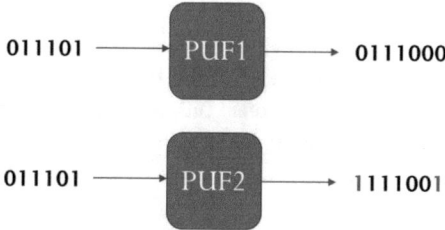

Fig. 2.18 An example of uniqueness evaluation of a PUF design

2.9.2   Reliability

It is a measure of the ability of the PUF to generate a consistent response R for a challenge C, regardless of any changes in the conditions of the environment such as the ambient temperatures and voltage supply. The reliability metrics is evaluated using the 'Intra-chip Hamming Distance'. If a single chip, represented as i, has the n-bit reference response $Ri(n)$ at normal operating conditions and the n-bit response $R'i(n)$ at different conditions for the same challenge C, the average intra-chip HD for k samples/chips is defined as [41]:

$$HD_{INTRA} = \frac{1}{k} \sum_{i=1}^{k} \frac{HD\left(R_i(n),\ R_i'(n)\right)}{n} \times 100\% \qquad (2.11)$$

From the intra-chip HD value, the reliability of a PUF can be defined as:

$$reliability = 100\% - HD_{INTRA} \qquad (2.12)$$

Take for example the PUF circuit depicted in Fig. 2.19. When a challenge (011101) is applied on this circuits at two different ambient temperatures, we ideally expect the hamming distance between the responses to be 0, however as can be seen from Fig. 2.19, in this case, the hamming distance is 1 (i.e. 10% difference from the original response), this indicates this PUF design is not a very reliable design.

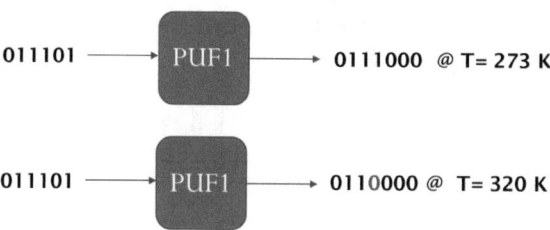

Fig. 2.19 An example of reliability evaluation of a PUF design

2.9.3   Uniformity

It is a measure of the "Unpredictability" of the PUF's responses, it is defined as the proportion of 0's and 1's in the response bits of a PUF, this percentage in a truly random response is 50%, it can be calculated using the average Hamming Weight of the responses as follows [41]:

$$Uniformity = \frac{1}{k} \sum_{i=1}^{k} r_i \times 100\% \qquad (2.13)$$

where k is the total number of responses and r_i is the Hamming Weight of the ith response

2.9.4   Tamper Resistance

It is a measure of how resistant a design is to tampering attempts, ideally the behaviour of the PUF should change completely if its design or structure are modified or "tampered with" in any way by adversaries. Mathematically, this can be expressed using 'Hamming Distance', between the responses from an authentic chip(i) and those of a tampered with chip(j), this can be calculated as follows:

$$HD_{AVE} = \frac{1}{CRP} \sum_{l=1}^{CRP} \frac{HD\big(R_i(l), R_j(l)\big)}{n} \times 100\% \qquad (2.14)$$

where:
  CRP is the total number of challenge/response pairs.
  $R_i(l)$ and $R_j(l)$ are the responses of the authentic and tampered with chip respectively for a specific challenge (l).
  If the PUF is tampered resistant against a specific physical modification, then the metric above should be 50% i.e. the modified PUF has a distinctly different behavior from that of the original PUF.

2.10   Case Study: FPGA Implementation of a Configurable Ring-Oscillator PUF

This section discusses the design process of a delay based PUF from the design rationale, through its detailed structure and silicon implementation, to its hardware testing and evaluation.

2.10.1   Design Rationale

The majority of systems based on configurable hardware platforms use third-party intellectual properties (IPs). A large percentage of such devices are not necessarily equipped with special on-board security features (e.g. trusted execution environment) due to cost limitations. For example, microcontrollers applications based on the ARM Cortex-M series does not support the ARM Trust Zone technology [42]. Such lightweight devices can be a prime target for adversaries aiming to extract intellectual properties. The use of PUF technology in such cases provides an affordable security solution. Ring oscillators PUFs have been proposed as the root of security in such applications, however; generic ring oscillator architectures such as described in Sect. 2.6.2 require significant logic resources (i.e. area overheads) in order to generate sufficient number of CRPs needed to ensure the protection of the PUF against emulating attacks and man-in the middle attacks [31]. In addition, existing solutions are not reproducible as they are developed for specific hardware platforms; moreover, they are based on manual routing techniques which are not well documented and hard to replicate [43–45].

The aim of this design is threefold:

(1)   To develop a configurable architecture which can be reprogramed in the field, such feature allow the client to re-configure the PUF in case they suspect its security has been compromised
(2)   To enhance the portability of RO-PUFs which facilitates their implementation on different hardware platforms
(3)   To increase the number of challenge/response pairs to better protect against emulation attacks.

2.10.2   Design Architecture

The top level architecture of the proposed design is the same as shown in Fig. 2.9 above, however its distinctive feature PUF is the re-configurable nature of the ring oscillator's chains which allows a significant increase in the number challenge-response pairs. The proposed PUF is composed of K ring oscillators, N multiplexers and a delay module. An implementation example is shown in Fig. 2.20, where in ($K = 4, N = 4$). The number of inverters in each ring oscillator (i.e. number of columns) is equal $N + 1$ (taking into consideration the NAND gate).

The operation principles of the above design are as follows: the challenge is applied at the SEL inputs; then the multiplexers select specific physical path for each ring oscillator, which generates two unique configurations for the ring oscillators, the outputs of which are compared using the counters and comparator logic. A unique response will be then generated. The delays of the physical paths of the two selected ring oscillators must be perfectly balanced, otherwise the PUF response will not be determined by the process variations-induced differences in the

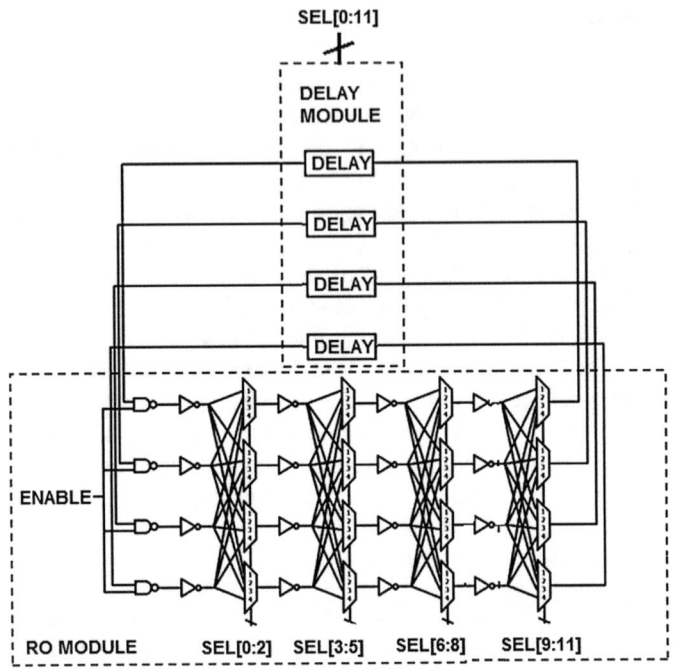

Fig. 2.20 Re-configurable ring oscillators chain

delay, but rather by the routing biases. This issue can be resolved by using hard macros and custom layout techniques as shown in [24, 45], however; these methods may be difficult to replicate and are only applicable to specific types of FPGAs architectures.

In this design, minimal routing constraints are imposed on the design, because symmetric routing requirements of inverter-chairs are satisfied using configurable delay elements, as shown in Fig. 2.20.

The use of such configurable architecture allows a significant increase in the number of CRPs, at every multiplexer stage two inverters are selected out of $(K.(K-1)/2)$ possible choices. For an N stage design, there will be $(K.(K-1))^N$ possible physical configurations of the two ring oscillators, whose outputs are to be compared. Each of these generates a unique response. Therefore, the number of challenge/response pairs that can be achieved is given in Eq. (2.15):

$$CRPs = \left(\frac{K.(K-1)}{2}\right)^N \qquad (2.15)$$

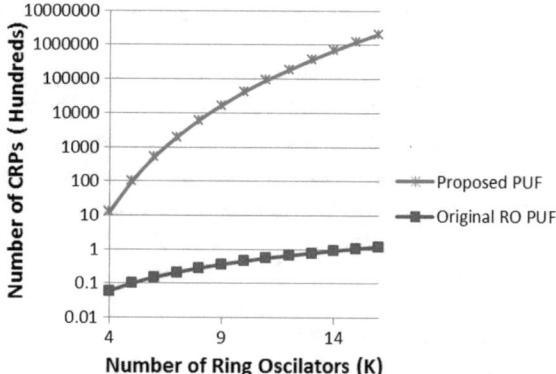

Fig. 2.21 The number of CRPs versus the number of ring oscillators (K) for the configurable RO PUF

This design offers a significantly larger number of *CRPs* compared to the traditional design in as shown in Fig. 2.21. The comparison was performed for a PUF with $N = 4$ stage, with different number of ring oscillators (i.e. K).

Figure 2.22 depicts the The Number of CRPs which can be obtained as a function of the number of Multiplexing Stages (N) for the Configurable RO PUF, it can be seen that for a 14 stage design, one can obtain more than one billion *CRPs*.

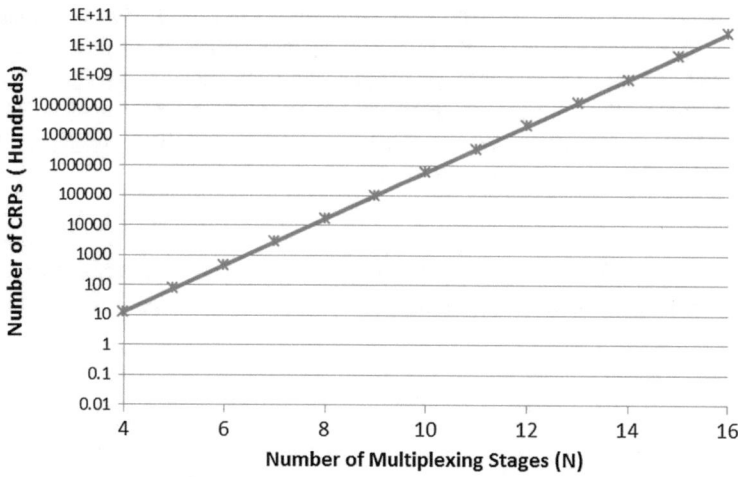

Fig. 2.22 The number of CRPs versus the number of multiplexing stages (*N*) for the configurable RO PUF

2.10.3 Hardware Implementation and Evaluation

To evaluate the proposed design, a proof of concept was implemented on an Altera Cyclone IV development board (EP4CE115F29C7), the choice of this device was mainly motivated by the ease of designing configurable delay element can be realized using this platform. The logic block in Altera FPGA is named as "Logic Array Block, each LAB consists of 16 Logic Elements (LEs) and each LE contains one LUT and register, this device contains a primitive called LCELL, which can be used as a buffer with a delay of 0.2 ns.

An example of the configurable delay module is shown in Fig. 2.23, where in it is possible to choose four different delay configurations by controlling the signal path between the input and the output using the two select bits.

A proof of concept design was implemented with four multiplexing stages ($N = 4$) and four chains of ring oscillators ($K = 4$) and their associated delays, each has 8 different delay configurations.

To assess the quality of those PUFs, we tested 9 devices. The PUFs were triggered using a separate controller module that repetitively applies the challenges and fetches the corresponding responses. For this experiment, only one hundred challenges have been applied (typically more CRPs are needed to properly assess the PUF).

To evaluate the Uniqueness metric, the Hamming Distances between responses from different devices for a chosen challenge have been calculated. The results depicted in Fig. 2.24, that the design under consideration has a Uniqueness of 48.9%, which is close to the ideal value, however the results also indicate that there are some correlation between devices with a worst case Hamming distance of around 20%. Ideally, the standard deviation of the Uniqueness graph should be very small, in other words, all measured Hamming Distances should be close to 50%, and this figure can be improved by increasing the number of ring oscillator chains, which allows for more variations between chips.

To evaluate the reliability metric, responses from the same device to a chosen challenge have been captured repeatedly, the average Hamming Distance between these responses is found to be 9.9% (see Fig. 2.25), this means that the design under consideration has a Reliability of around 91%. Although the ideal value is 100% for

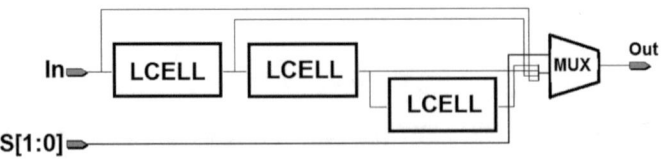

Fig. 2.23 A configurable delay module using LCELL

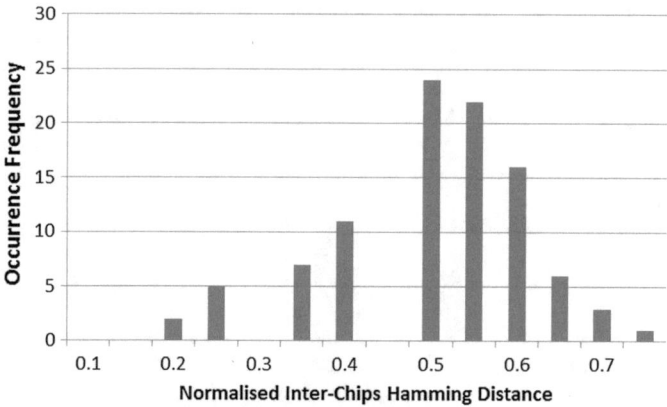

Fig. 2.24 The evaluation of the PUF uniqueness using normalized inter-chips hamming distance

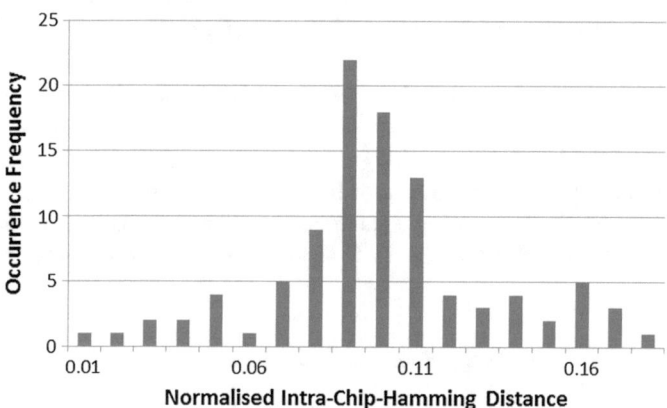

Fig. 2.25 The evaluation of the PUF reliability using normalized intra-chip hamming distance

this metric, standard error correction algorithms can easily enhance improve this figure, the use of such algorithms to enhance the reliability of PUF designs will be discussed in a greater depth in Chap. 4.

To evaluate the uniformity metric, the Hamming Weight of responses from the same device to different challenge has been estimated. The results, depicted in Fig. 2.26, indicate that the design under consideration has a Uniformity of around 53%, which is very close to the ideal value, which indicate that this design is highly unpredictable, hence more resilient to machine learning modelling attacks.

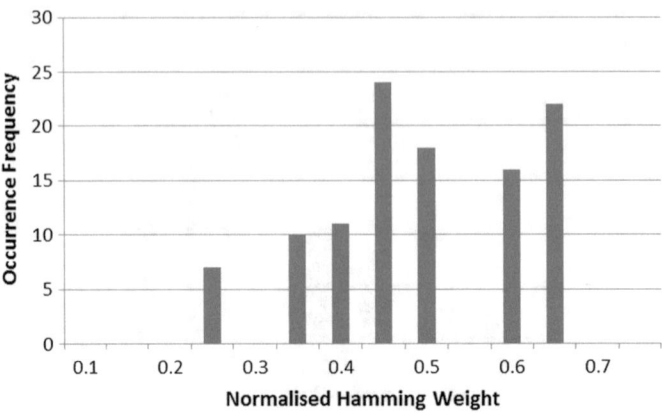

Fig. 2.26 The evaluation of the PUF uniformity using normalized hamming weight

2.11 Comparison of Existing PUFs Design

This section provides a comparison of the performance figures of publically available silicon implementations of physically unclonable functions, these are summarised in Table 2.1. From a designer perspective, the choice of a suitable PUF is driven solely by the requirements of intended applications. Let us consider the case of using PUFs for on chip generation of encryption keys. In this type of applications, the reliability of the PUF is the most important metric in order to

Table 2.1 A comparison of the performance figures of silicon implementations of physically unclonable functions

PUF name	Hybrid [49]	PTAT [50]	osc collapse [46]	INV_PUF [47]	SRAM [47]	RO [47]
Type	Voltage (Metastability)	Voltage	Delay	Voltage (Monostable)	Voltage (Memory)	Delay
Technology (nm)	22	65	40	65	65	65
Nominal voltage (V)	0.8	1	0.9	1	1	0.5
Energy per bit (pJ/bit)	0.19	1.1	17.75	0.015	1.1	0.475
Area (μm^2)	24,000	3.07	527,850	6000	48,705	102,900
Uniqueness	0.49	0.5001	0.5007	0.501	0.33	0.473
Reliability	0.026	0.0057	0.01	0.003	0.06	0.045
CRPs	196	1	5.50E + 28	11	512	1024

ensure that the PUF is able to consistently generate the same response for a chosen challenge, so in this case the osc collapse [46] in Table 2.1 is the best bet, as it has the most stable response.

On the other hand, if the PUF is to be used for the authentication of energy–constrained devices, then uniqueness, reliability, and energy efficiency need to be considered, so in this case INV_PUF [47] can provide an acceptable trade-off, as it has the lowest energy consumption and an acceptable reliability figure. More comparison data of existing implementations of PUF architectures can be found in [48].

2.12 Conclusions

Physical disorder is an omnipresent phenomenon which can be observed at the Nona-scale level in the form of irregular structures of physical objects. The continuous scaling of semiconductor technology has made it impossible to have precise control of the dimensions of fabricated devices, this is known as process variations, and it can be seen as another manifestation of physical disorder wherein each implementation of a circuit has unique characteristics. Silicon-based physically unclonable devices are integrated circuits which exploit this phenomenon to generate a unique digital identifier for each chip. There are great many techniques to design a physically unclonable function, but they all share the same principles:

- A PUF circuit typically consists of two stages, the first stage converts inherent process variation to a measurable quantity (e.g. voltage, current and delay), the latter is then converted to a binary response in the second stage.
- The technology with which the PUF circuit is to be fabricated, needs to have sufficient process variations, otherwise the difference between silicon chips is going to be too small to generate unique identifiers.
- The PUF circuit should generate consistent responses under different environment conditions; this is measured using the reliability metric, which is calculated using the average Hamming distance between responses from the same chip to a chosen challenge under various environment conditions. Ideal value for the reliability metric is "0".
- The PUF circuit should generate a unique response for each chip for the same challenge; this is measured using the uniqueness metric which is calculated using the average Hamming Distance between responses from the different chip to the same challenge. Ideal value for the Uniqueness metric is "0.5".
- The PUF circuit should produce unpredictable responses; one of the indicators of unpredictability is the average Hamming weight of its responses that should ideally be 0.5.

2.13   Problems

2.1. Figure 2.27 shows a synchronous Arbiter PUF circuit, where in the arbiter
  block is realised using the structure shown in Fig. 2.7. This PUF design is
  implemented in 130 nm technology where $T = 66$ ps. Due to the very small
  difference in the delay of the two inverter chains, a metastability event occurs
  every time a challenge is applied to the input of this circuit. You are given the
  following information; the clock frequency is 100 MHz, the voltage at the
  output of the gate G1 (in Fig. 2.7) in the arbiter block is ($V_0 = 0.2v$), and its
  threshold voltage is ($V_{th} = 0.6v$).

  1. Calculate the probability of a metastability induced error at the output of
     the PUF circuit.
  2. Can this error rate be reduced by changing the clock frequency?

2.2. How many stages of an arbiter PUF based on the structure shown in Fig. 2.8,
  do you need to obtain a number of challenge/response pairs equivalent to that
  of a generic RO PUF (see Fig. 2.9) which has 10 ring oscillators.

2.3. Self-timed rings may not oscillate unless their initial state is setup correctly.
  One way to avoid a deadlock state in the STR PUF design is to store valid
  initial states as bit patterns in is an on-chip non-volatile memory; these can be
  used to initiate the oscillation. What are the bit patterns a designer would need
  to store to ensure the oscillation of a four stage self-timed ring that has the
  same structure as the one described in Fig. 2.11?

2.4. A two-choose-one (TCO) PUF is designed to have 512 challenge/response
  pairs, assuming the size of its comparator circuit is equivalent to 6 transistors.

  1. Estimate the area of this PUF in terms of transistor counts.
  2. How does this compare to the size of a single-bit response SRAM memory
     with the same number of challenge/response pairs?

2.5. A DRAM memory, which has 64 cells arranged in 16 rows and 4 columns, is
  used to design a physically unclonable function as described in Sect. 2.7,
  where the size of the reserved region is 12 cells. Assuming reserved regions
  can only be defined using three adjacent rows:

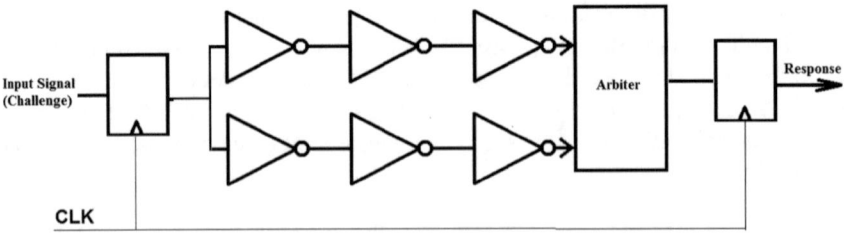

Fig. 2.27 A synchronous arbiter PUF circuit

Table 2.2 Challenge/response behaviour for a PUF design

(a) For different devices

Challenge	Device			
	Chip 1	Chip 2	Chip 3	Chip 4
000	11010110	01100001	10001001	11101101
001	00011100	00010000	11001010	01010101
010	10110100	11000111	01010100	11000011
011	10000101	10000101	10111011	01111111
100	11010011	11001010	01010100	11101001
101	10110101	11101000	11110101	11001111
110	11000101	01000011	01001000	11011000
111	11000101	01110101	01100001	01000101

(b) At different temperatures

Challenge	Device		
	Chip 1 (T = 270 K)	Chip 1 (T = 300 K)	Chip 1 (T = 320)
000	11010000	11010001	11010001
001	11000010	11000000	11000010
010	10011000	10011001	10011001
011	11000000	11000001	11000001
100	10000111	10000110	10000111
101	11111110	11111111	11111111
110	10000111	10000110	10000110
111	11101001	11101000	11101001

1. How many bits does the response of this PUF have?
2. Calculate the total number of challenge/response pairs assuming each reserved DRAM region exhibits four distinct decay behavior ($N_T = 4$) during the allocated decay time.

2.6. A PUF circuit, which has 3-bits challenge and 5-bits responses, is implemented on four chips. The challenge/response behaviour of this is design is summarised in Table 2.2.

1. Calculate the Uniqueness metric of this PUF
2. Calculate the reliability metric under temperature variations
3. Calculate the Uniformity metric of this design.

2.7. A PUF circuit, which has 3-bits challenge and 5-bits responses, was implemented and given to Alice to use as an access authentication device. Steve, a known adversary of Alice, managed to gain access and maliciously changed the PUF circuit to insert a small passive component, which works as an antenna, this has led to some changes of the behaviour of the original PUF as

Table 2.3 Challenge/response behaviour for authentic versus "Tampered with" PUF design

Challenge	Device	
	Original chip	"Tampered with" chip
000	11010000	11010001
001	11000010	11000000
010	10011000	10011001
011	11000000	11101001
100	10000111	11001111
101	11111110	11011000
110	10000111	01000101
111	11101001	11101001

summarised in Table 2.3. Analyse the data provided below and explain whether or not Alice's PUF can be considered resistant to Steve's tampering attack.

References

1. L. Daihyun, J.W. Lee, B. Gassend, G.E. Suh, M.V. Dijk, S. Devadas, Extracting secret keys from integrated circuits. IEEE Trans. Very Large Scale Integr. VLSI Syst. **13**, 1200–1205 (2005)
2. T. Fournel, M. Hébert, Towards weak optical PUFs by random spectral mixing, in *2016 15th Workshop on Information Optics (WIO)* (2016), pp. 1–3
3. S. Dolev, L. Krzywiecki, N. Panwar, M. Segal, Optical PUF for non forwardable vehicle authentication, in *2015 IEEE 14th International Symposium on Network Computing and Applications* (2015), pp. 204–207
4. D.R. Reising, M.A. Temple, J.A. Jackson, Authorized and Rogue device discrimination using dimensionally reduced RF-DNA fingerprints. IEEE Trans. Inf. Forensics Secur. **10**, 1180–1192 (2015)
5. M.W. Lukacs, A.J. Zeqolari, P.J. Collins, M.A. Temple, RF-DNA fingerprinting for antenna classification. IEEE Antennas Wirel. Propag. Lett. **14**, 1455–1458 (2015)
6. W.E. Cobb, E.D. Laspe, R.O. Baldwin, M.A. Temple, Y.C. Kim, Intrinsic physical-layer authentication of integrated circuits. IEEE Trans. Inf. Forensics Secur. **7**, 14–24 (2012)
7. J.D.R. Buchanan, R.P. Cowburn, A.-V. Jausovec, D. Petit, P. Seem, G. Xiong et al., Forgery:/ 'Fingerprinting/' documents and packaging. *Nature* **436**, 475–475 (2005), 07/28/print
8. S. Nassif, K. Bernstein, D.J. Frank, A. Gattiker, W. Haensch, B.L. Ji et al., High performance CMOS variability in the 65 nm regime and beyond, in *IEEE International Electron Devices Meeting* (2007), pp. 569–571
9. A. Narasimhan, R. Sridhar, Impact of variability on clock skew in H-tree clock networks, in *International Symposium on Quality Electronic Design* (2007), pp. 458–466
10. S.R. Nassif, Modeling and analysis of manufacturing variations, in *IEEE Conference on Custom Integrated Circuits* (2001), pp. 223–228
11. A. Chandrakasan, W.J. Bowhill, *Design of High-Performance Microprocessor Circuits* (Wiley-IEEE Press, 2000)
12. International Technology Roadmap for Semiconductors (www.itrs.net).

13. B.P. Wong, A. Mittal, Z. Gau, G. Starr, *Nano-CMOS Circuits and Physical Design* (Wiley, Hoboken, New Jersey, 2005)
14. D. Marculescu, S. Nassif, Design variability: challenges and solutions at microarchitecture-architecture level, in *Design, Automation and Test Conference in Europe* (2008)
15. Wikipedia. Available: https://en.wikipedia.org
16. S. Graybeal, P. McFate, Getting out of the STARTing block. *Sci. Am.* **261** (1989)
17. B. Halak, A. Yakovlev, Fault-tolerant techniques to minimize the impact of crosstalk on phase encoded communication channels. Comput. IEEE Trans. **57**, 505–519 (2008)
18. I.M. Nawi, B. Halak, M. Zwolinski, The influence of hysteresis voltage on single event transients in a 65 nm CMOS high speed comparator, in *2016 21th IEEE European Test Symposium (ETS)* (2016), pp. 1–2
19. I.M. Nawi, B. Halak, M. Zwolinski, Reliability analysis of comparators, in *2015 11th Conference on Ph.D. Research in Microelectronics and Electronics (PRIME)* (2015), pp. 9–12
20. Y. Zhang, P. Wang, G. Li, H. Qian, X. Zheng, Design of power-up and arbiter hybrid physical unclonable functions in 65 nm CMOS, in *2015 IEEE 11th International Conference on ASIC (ASICON)* (2015), pp. 1–4
21. L. Lin, S. Srivathsa, D.K. Krishnappa, P. Shabadi, W. Burleson, Design and validation of arbiter-based PUFs for sub-45-nm low-power security applications. IEEE Trans. Inf. Forensics Secur. **7**, 1394–1403 (2012)
22. Y. Hori, T. Yoshida, T. Katashita, A. Satoh, Quantitative and statistical performance evaluation of arbiter physical unclonable functions on FPGAs, in *2010 International Conference on Reconfigurable Computing and FPGAs* (2010), pp. 298–303
23. C.E.D. Yin, G. Qu, LISA: maximizing RO PUF's secret extraction, in *2010 IEEE International Symposium on Hardware-Oriented Security and Trust (HOST)* (2010), pp. 100–105
24. X. Xin, J.P. Kaps, K. Gaj, A configurable ring-oscillator-based PUF for Xilinx FPGAs, in *2011 14th Euromicro Conference on Digital System Design* (2011), pp. 651–657
25. S. Eiroa, I. Baturone, An analysis of ring oscillator PUF behavior on FPGAs, in *2011 International Conference on Field-Programmable Technology* (2011), pp. 1–4
26. S. Eiroa, I. Baturone, Circuit authentication based on Ring-Oscillator PUFs, in *2011 18th IEEE International Conference on Electronics, Circuits, and Systems* (2011), pp. 691–694
27. H. Yu, P.H.W. Leong, Q. Xu, An FPGA chip identification generator using configurable ring oscillators. IEEE Trans. Very Large Scale Integr. VLSI Syst. **20**, 2198–2207 (2012)
28. M. Delavar, S. Mirzakuchaki, J. Mohajeri, A Ring Oscillator-based PUF with enhanced challenge-response pairs. Can. J. Electr. Comput. Eng. **39**, 174–180 (2016)
29. A. Cherkaoui, L. Bossuet, C. Marchand, Design, evaluation, and optimization of physical unclonable functions based on transient effect ring oscillators. IEEE Trans. Inf. Forensics Secur. **11**, 1291–1305 (2016)
30. L. Bossuet, X.T. Ngo, Z. Cherif, V. Fischer, A PUF based on a transient effect ring oscillator and insensitive to locking phenomenon. IEEE Trans. Emerg. Top. Comput. **2**, 30–36 (2014)
31. B. Halak, Y. Hu, M.S. Mispan, Area efficient configurable physical unclonable functions for FPGAs identification, in *2015 IEEE International Symposium on Circuits and Systems (ISCAS)* (2015), pp. 946–949
32. D.J. Kinniment, *Synchronization and Arbitration in Digital Systems* (2007)
33. G.E. Suh, S. Devadas, Physical unclonable functions for device authentication and secret key generation, in *2007 44th ACM/IEEE Design Automation Conference* (2007), pp. 9–14
34. J. Murphy, M.O. Neill, F. Burns, A. Bystrov, A. Yakovlev, B. Halak, Self-timed physically unclonable functions, in *2012 5th International Conference on New Technologies, Mobility and Security (NTMS)* (2012), pp. 1–5

35. M. Kalyanaraman, M. Orshansky, Novel strong PUF based on nonlinearity of MOSFET subthreshold operation, in *2013 IEEE International Symposium on Hardware-Oriented Security and Trust (HOST)* (2013), pp. 13–18
36. M.S. Mispan, B. Halak, Z. Chen, M. Zwolinski, TCO-PUF: a subthreshold physical unclonable function, in *Ph.D. Research in Microelectronics and Electronics (PRIME), 2015 11th Conference on* (2015), pp. 105–108
37. W. Xiong, A. Schaller, N.A. Anagnostopoulos, M.U. Saleem, S. Gabmeyer, S. Katzenbeisser, J. Szefer, Run-time accessible dram pufs in commodity devices, in *Presented at the Cryptographic Hardware and Embedded Systems International Conference* (Santa Barbara, CA, USA, 2016)
38. S.S.K.J. Guajardo, G.-J. Schrijen, P. Tuyls, FPGA intrinsic PUFs and their use for IP protection, in *International Conference on Cryptographic Hardware and Embedded Systems* (2007), pp. 63–80
39. N. Torii, D. Yamamoto, T. Matsumoto, Evaluation of latch-based PUFs implemented on 40 nm ASICs, in *2016 Fourth International Symposium on Computing and Networking (CANDAR)* (2016), pp. 642–648
40. A. Stanciu, M.N. Cirstea, F.D. Moldoveanu, Analysis and evaluation of PUF-based SoC designs for security applications. IEEE Trans. Industr. Electron. **63**, 5699–5708 (2016)
41. V.G.A. Maiti, P. Schaumont, A systematic method to evaluate and compare the performance of physical unclonable functions. IACR ePrint **657**, 245–267 (2013)
42. (2017). *ARM TrustZone Overview*. Available: https://www.arm.com/products/security-on-arm/trustzone
43. J. Zhang, Q. Wu, Y. Lyu, Q. Zhou, Y. Cai, Y. Lin et al., Design and implementation of a delay-based PUF for FPGA IP protection, in *2013 International Conference on Computer-Aided Design and Computer Graphics* (2013), pp. 107–114
44. F. Kodýtek, R. Lórencz, A design of ring oscillator based PUF on FPGA, in *2015 IEEE 18th International Symposium on Design and Diagnostics of Electronic Circuits & Systems* (2015), pp. 37–42
45. L. Feiten, T. Martin, M. Sauer, B. Becker, Improving RO-PUF quality on FPGAs by incorporating design-dependent frequency biases, in *2015 20th IEEE European Test Symposium (ETS)* (2015), pp. 1–6
46. K. Yang, Q. Dong, D. Blaauw, D. Sylvester, A physically unclonable function with BER smaller than 10e-8 for robust chip authentication using oscillator collapse in 40 nm CMOS, in *2015 IEEE International Solid-State Circuits Conference—(ISSCC) Digest of Technical Papers* (2015), pp. 1–3
47. A.B. Alvarez, W. Zhao, M. Alioto, Static physically unclonable functions for secure chip identification with 1.9–5.8% native bit instability in 65 nm. IEEE J. Solid-State Circuits **51**, 763–775 (2016)
48. M. Alioto, A. Alvarez, *Physically Unclonable Function database*. Available: http://www.greenic.org/pufdb
49. S.K. Mathew, S.K. Satpathy, M.A. Anders, H. Kaul, S.K. Hsu, A. Agarwal et al., 16.2 A 0.19pJ/b PVT-variation-tolerant hybrid physically unclonable function circuit for 100% stable secure key generation in 22 nm CMOS, in *2014 IEEE International Solid-State Circuits Conference Digest of Technical Papers (ISSCC)* (2014), pp. 278–279
50. J. Li, M. Seok, A 3.07 square micron bitcell physically unclonable function with 3.5 and 1% bit-instability across 0–80 C and 0.6–1.2 V in a 65 nm CMOS, in *2015 Symposium on VLSI Circuits (VLSI Circuits)* (2015), pp. C250–C251

Reliability Challenges of Silicon-Based Physically Unclonable Functions

3

3.1 Introduction

The continuous scaling of semiconductor technologies and their associated manufacturing variations has made it possible to build silicon-based physically unclonable functions, a new cryptographic primitive capable of generating unique hardware identifiers. Scaling, however, has given rise to major reliability challenges.

New technologies are considered less mature than older ones; this is partly because they require the use of new materials and process steps that are not well characterised before being deployed in commercial products. What is more, smaller feature sizes and the increase in electric fields and operating temperatures have made modern integrated circuits more prone to reliability problems, which used to be seen as second order effects in the past. For example, soft errors induced by the propagation of single event transients have become a significant reliability problem in sub-100 nm CMOS chips [1]. Aging is another failure mechanism affecting modern systems, wherein, a progressive degradation of electrical characteristics of transistors and wires takes place, which affects circuits' performance and may cause functional errors [2].

It is important that these reliability problems are effectively resolved when building silicon-based physically unclonable functions; this is because the applications of the PUF technology, such as cryptographic key generation and devices authentication, rely on the fact that these primitives are capable of generating consistent responses for the corresponding challenges.

This chapter aims to:

1. Explain the physical origins of the major reliability issues affecting CMOS technology.
2. Discuss how these issues can affect the usability of PUF technology.
3. Present a case study on the evaluation of the impact of CMOS aging on the quality metrics of PUF designs.

© Springer International Publishing AG, part of Springer Nature 2018
B. Halak, *Physically Unclonable Functions*,
https://doi.org/10.1007/978-3-319-76804-5_3

It is hoped that this chapter will give the reader an in-depth understanding of the reliability challenges in nano-scale CMOS technologies and their effects on the usability of silicon-based PUF designs.

3.2 Chapter Overview

The organisation of this chapter is as follows, Sect. 3.3 explains in depth the physical mechanisms of CMOS aging, and discusses the susceptibility of PUF designs to the different forms of aging-related circuits' degradation. Section 3.4 explains the causes of typical temporal failure mechanisms, including radiation hits, electromagnetic interference, thermal noise and ground bounces, this is followed by a discussion on how these phenomena affect the different types of PUF architectures. Section 3.5 presents a case study to illustrate how the quality metrics of a PUF design can be assessed given a specific reliability challenge, the study focuses on the impact of BTI aging on three different PUF architectures, namely: SRAM, TCO and Arbiter. Conclusions and lessons learned are presented in Sect. 3.6. Finally, a set of problems and exercises are included in Sect. 3.7.

3.3 Aging Issues in CMOS Circuits

Aggressive technology scaling below 32 nm has aggravated the aging problem of CMOS devices. The latter phenomenon is characterised by a progressive degradation of the performance and reliability of devices and system components. It is caused by several mechanisms, namely: bias temperature instability (BTI), hot carrier injection (HCI), time-dependent dielectric breakdown (TDDB) and electromigration [3, 4]. This section explains the physical causes for aging and summarises how it affects the quality metrics of a PUF design.

3.3.1 Bias Temperature Instability (BTI)

BTI is usually considered as the most concerning aging mechanism in modern technologies [5]. This mechanism is characterised by a positive shift in the absolute value of the threshold voltage of the MOSFET (metal–oxide–semiconductor field-effect transistor) device. Such a shift is typically attributed to hole trapping in the dielectric bulk and the breakage of Si–H bonds at the Si-dielectric interface [6]. Negative bias temperature instability (NBTI) in PMOS transistors prevails over positive bias temperature instability (PBTI) exhibited by NMOS devices [7]. However, with the use of high-K dielectric stacks, PBTI is becoming more significant and can no longer be neglected.

Reaction–diffusion (R–D) model is a well-known model to explain BTI effects [8]. According to this model, BTI mechanism is divided into two critical steps: reaction and diffusion. During the reaction phase, Si–H bonds are disassociated under stress; this produces hydrogen and traps at oxide interface. These hydrogen atoms or molecules diffuse into the interface at diffusion phase. Let us take NBTI as an example, in this case, traps are generated and hydrogen is diffused toward gate when the PMOS transistor is negatively biased ($Vgs = -Vdd$). Once the stress is removed (e.g. the negative biasing is taken away), some hydrogen diffuses back and recombines with traps to recover the bonds. Figure 3.1 demonstrates these stresses and recovery phases.

The reaction–diffusion model, proposed in [8], allows designers to estimate the drift of threshold voltage (ΔV_{th}) induced by BTI aging as a function of technology's parameters, operating conditions and time. A closed form analytical model has

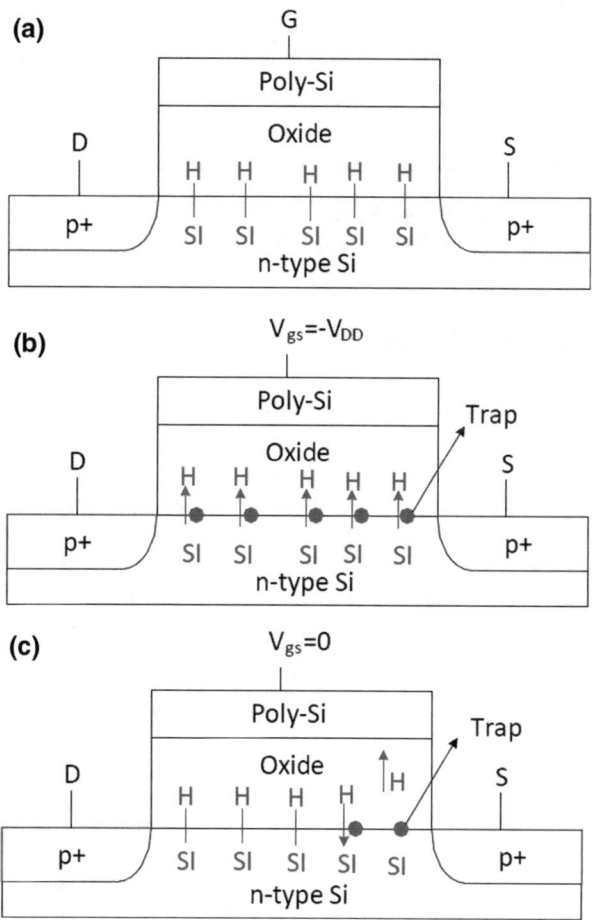

Fig. 3.1 NBTI Aging based on R–D mechanism: **a** fresh PMOS, **b** stress phase, **c** recovery phase

recently been proposed, which allows designers to estimate long-term, worst-case threshold voltage shift as follows:[9, 10]:

$$\Delta V_{th} = \chi K \sqrt{C_{ox}(Vdd - V_{th})} \exp\left(\frac{E_a}{k_B T_A}\right)(\alpha t)^{1/6}, \tag{3.1}$$

where

C_{ox} is the oxide capacitance.

t   is the operating time.

α is the fraction of the operating time during which a MOS transistor is under a stress condition. It has a value between 0 and 1. $\alpha = 0$ if the MOS transistor is always OFF (recovery phase), while $\alpha = 1$ if it is always ON (stress phase).

E_a is the activation energy ($E_a \cong 0.1$ eV).

k_B is the Boltzmann constant.

T_A is the aging temperature.

χ is a coefficient to distinguish between PBTI and NBTI. Particularly, χ equals 0.5 for PBTI, and 1 for NBTI.

K Lumps technology specific and environmental parameters.

K has been estimated to be 2.7 $V^{1/2}F^{-1/2}s^{-1/6}$ by fitting the model with the experimental results reported in [7] for a 32 nm high-k CMOS technology.

3.3.2 Hot Carrier Injection (HCI)

Aging induced by the HCI mechanism is caused by the generation of traps in the gate oxide. These are created when channel current carriers are accelerated to a sufficient energy to cross into oxide. Thus, it mostly happens when transistor is pinched off [11]. Additionally, HCI mainly occurs in NMOS transistors as opposed to PMOS since electrons obtain a greater mobility than holes, thereby could speed up to a higher energy. Unlike BTI, the traps produced by HCI are permanent, therefore, the recovery phase is negligible. Figure 3.2 illustrates the HCI mechanism in an NMOS transistor.

Fig. 3.2 HCI aging of an NMOS transistor

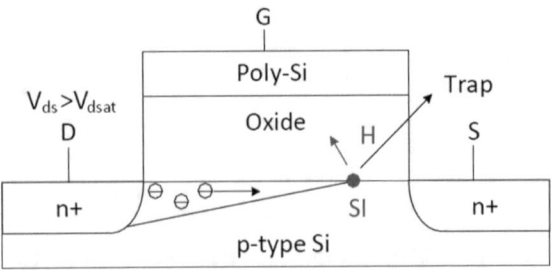

This mechanism is characterised by a shift in the absolute value of the threshold voltage (ΔV_{th}) of the NMOS device which is proportional to the number of traps ($\Delta N_{it}(t)$). The "lucky electron" model is used to estimate the damage inflicted on the device, in terms of the generated interface traps ($\Delta N_{it}(t)$) [11].

$$\Delta N_{it}(t) = C\left[\frac{I_{ds}}{W} \, exp\left(-\frac{\emptyset_{it,e}}{q\lambda_e E}\right)t\right]^n \qquad (3.2)$$

where

c	is a process-dependent constant.
I_{ds}	is the drain-to-source current.
W	is the transistor width.
$\emptyset_{it,e}$	is the critical energy needed to overcome oxide interface barrier.
E	is the electric field at the drain.
λ_e	is the hot-electron mean-free path.
t	is the fraction of the operating time during which a MOS transistor is under a stress condition.

At the initial stage of the of interface-trap generation process, the rate is reaction-limited, therefore, Nit(t) $\propto$ t and n = 1; at later stage, the generation is diffusion limited (n = 0.5).

3.3.3 Time-Dependent Dielectric Breakdown (TDDB)

Voltage drops across the gate oxide of CMOS transistors lead to the creation of traps within the dielectrics, such defects may eventually join together to form a conductive path through the stack which causes an oxide breakdown, hence a device failure [12]. This mechanism is especially a concern in the technologies of the nanometre range, because of the reduction of the critical density of traps needed to form a conducting path through these thin layers, in combination with stronger electric field across the dielectrics.

3.3.4 Electromigration

Electromigration refers to the displacement of metal atoms in the wires caused by the transfer of momentum from the conducting electrons, particularly in high current density interconnect and in visas. This electron wind creates hillocks and voids as shown in Fig. 3.3. The former can cause shorts between neighbouring wires and the latter causes opens, both of which may cause functional errors.

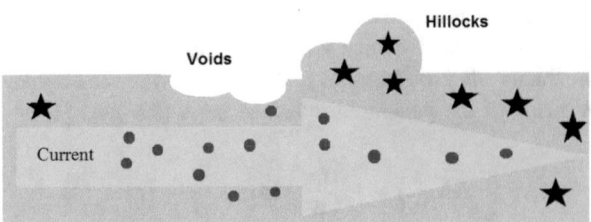

Fig. 3.3 The generation of hillocks and voids in metal conductors

The decreasing dimensions of interconnect combined with the increase in the operating frequencies in nano-scale CMOS technologies amplifies the effects of electromigration, especially for technologies beyond 28 nm [13–15].

3.3.5   On the Susceptibility of PUFs to Aging Mechanisms

The progressive degradation of the performance and reliability of CMOS devices caused by the above aging mechanisms does certainly affect the quality metrics of PUF circuits. The magnitude of such impact very much depends on the type of PUF designs. For example; BTI is the most likely cause of aging in a stand alone SRAM-based PUFs; this is because such devices normally store the same data for long periods of time, which means some of their transistors, may be under constant BTI stress conditions. On the other hand, electromigration issues are unlikely to cause major reliability problems due to the relatively short period of activities in such PUFs. It should be noted that powering down the whole SRAM block can mitigate against aging, but may not always be possible.

Another example is ring oscillator PUF designs, wherein the inverter chains are subjected to periodic intervals with high level of switching activities. Therefore, in these designs, hot electron injection and electromigration can have significant aging impact. Overall, the impact of CMOS aging should be given due consideration at the design stage in order to ensure reliable operation of PUF devices over the expected lifetime of their associated products.

3.4   Sources of Soft Errors in CMOS Integrated Circuits

The scaling of semiconductor technologies has increased the vulnerability of integrated circuits to soft errors. The latter typically occur at random and last for a short period. There are two types of soft errors, *transient faults* that are caused by environmental conditions, e.g. random bombardment by ionising particles, and intermittent faults that are caused by non-environmental conditions (e.g. loose connections, crosstalk glitches, ground bounces and instantaneous IR drops in voltage supply).

This section explains the physical mechanisms of the above phenomena.

3.4.1 Radiation-Induced Soft Errors

Soft errors caused by radiation-induced transient glitches in electronic systems are another reliability concern for integrated circuit designers. There are two major sources of radiation, namely: neutrons from cosmic rays [16] and alpha particles from the packaging and bonding materials [17]. Although the impact on electronic devices is more pronounced in aerospace and avionic applications, faults induced in commercial and consumer electronics are frequently reported especially in devices which operate at high altitudes [18].

The alpha particles caused by atmospheric neutrons are the main source of radiation-induced soft errors in electronic devices at the ground level.

An alpha particle can generate around one million pairs of electron and holes if it hits a sensitive area in a device such as a transistor drain or a depletion region. This may lead a transient voltage pulse when these free carriers are driven by the electric field as shown in Fig. 3.4.

Such a phenomenon is especially problematic for dense memory arrays (SRAM and DRAM) and sequential logic (Flip-flops and latches) as it may lead to reversing their original state, hence inducing an error in their respective contents [19, 20].

A radiation hit to the combinational logic can also lead to an error if the induced transient pulse propagates through the logic chain and reaches the storage element during its latching window [21].

Techniques to mitigate the effects of such errors on memory blocks mainly consist of using error correction codes [22]. On the other hand, radiation-hardening methods are typically employed to protect storage elements and combinational logic against such errors [23].

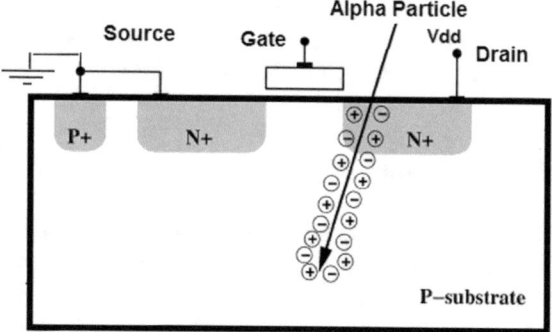

Fig. 3.4 The generation of free carriers by an alpha particle hit on an NMOS device

3.4.2   Crosstalk Noise

This is one of the major reliability issues that especially affect the integrity of on-chip communication schemes. It is defined as an undesirable electromagnetic coupling between switching and non-switching signal lines [24]. Failure mechanisms caused by crosstalk in on-chip communication links can be classified into two types as follows [25]:

(a)   Functional failures are caused by static noise. The latter is defined as the noise pulse induced on a quiet victim net due to switching of neighbouring aggressors. If this pulse propagates through logic gates and reaches storage elements, it may lead to sampling the wrong data.
(b)   Timing Failures occur when the delay of a communication bus increases, which leads to a timing violation, thus an error. The increase in the delay is caused by a change in the coupling capacitance between adjacent wires due to the switching activity of neighbouring nets (i.e. *Miller effect*).

These failure mechanisms can be the cause of faults in communication links, hence system malfunction [24–33]. In addition, single event upsets (SEU), caused by random bombardment of integrated circuits by ionising particles of a very high energy, may lead to a rapid change in node voltage. These changes induce crosstalk pulses in adjacent interconnects, which spreads a single event transient (SET) pulse resulting from the hit node to other parts of the circuit. This phenomenon, caused by capacitive coupling, has also been reported in the literature [34], the authors concluded that the effects of single event transients caused by crosstalk increases as devices scale down, as the energy of the ionising particle increases and as the interconnect lengths increases.

Crosstalk may also cause permeant device failure in some cases. Consider a victim and aggressor wires, which are driven by a static "1" and a fast-rising transition (from 0 to 1), respectively. If the height of the generated crosstalk glitch is larger than the upper value of logic-high voltage, the crosstalk glitch will create reliability problem of the device in the end. This is because excessive over-voltage at the gate input reduces the life expectancy of the device drastically as the gate oxide layer gradually suffers from time-dependent-dielectric breakdown and hot-carrier effect as described above [35].

3.4.3   Ground Bounce

This is a phenomenon that is typically induced by a sudden change in the current drawn from the voltage supply such as the case at the rising edge of the clock in synchronous systems [36] or when the power gating structures are activated/deactivated during periods of simultaneous switching activities on chips [37]. In such cases, voltage spikes may develop across the power/ground parasitic inductances when the switching takes place. These glitches can then propagate to functional blocks on chip through the power or clock distribution networks and causes soft errors.

3.4.4 Thermal Noise

It refers to the voltage fluctuations caused by the random thermally-induced motion of electrons in a resistive medium such as polysilicon resistors and the channel of MOSFET transistors [38]. It has a very wide bandwidth with a zero mean. It can be modelled as a voltage source (v) series with a resistor (R) with a power spectral density given as follows:

$$\overline{v^2} = 4k_B TR \tag{3.3}$$

$\overline{v^2}$ The voltage variance (mean square) per hertz of bandwidth.

k_B Boltzmann's constant in joules per kelvin.

T The resistor's absolute temperature in kelvin.

R The resistance of the medium in ohms (Ω).

This type of noise is particularly problematic in three-dimensional integrated circuits (3-D IC) [39], this is because the internal layers in those chips are detached from the heat sink, in addition, the transfer of heat is further restricted by interlayer dielectric and oxide-based bonding layers with low thermal conductivity. This heating effect leads to an increase in the magnitude of the thermal noise, which in turn degrades the integrity of signals.

3.4.5 On the Susceptibility of PUFs to Soft Errors

The advances in the fabrication process of CMOS technologies, which allow the construction of densely integrated circuits, have aggravated some of its inherent reliability challenges. This may lead to degradation in the quality metrics of silicon-based unclonable functions. The magnitude of such degradation and how detrimental it is going to be to the usability of the PUF is very much dependent on the nature of the systems incorporating the PUF circuit. For example, in high-speed 3D-IC chips, soft errors induced by electromagnetic interference and thermal noise are likely to be the largest source of soft errors. On the other hand, in low power embedded systems, electromagnetic noise may be less of an issue compared to ground bounce induced by switching activities of power gating structures. The type of the PUF design also greatly affects its susceptibility to certain physical mechanisms of soft errors. For example, an Arbiter PUF is generally more vulnerable to crosstalk-induced errors compared to memory-based designs (e.g. SRAM PUF). This is because temporal electromagnetic coupling can increase the delay of one of logic paths of an arbiter PUF, which may cause the latter to produce a different response from what it would have generated without crosstalk noise.

In the case of SRAM PUFs, which is based on the use of start-up values of the cells when the memory is powered up, electromagnetic interference is less likely to be a major cause of soft errors.

Soft errors may undermine the PUF's capability of generating consistent responses, which may prove detrimental to its usability for the generation of cryptographic keys, which require zero bit error rates. Other applications of the PUF technology such as authentication schemes may be more tolerant to such transient issues.

There are a number of measures to mitigate the impact of soft errors such as crosstalk avoidance methods [40], radiation hardened design [23] and error correction schemes. Designers need to take into consideration the structure of the PUF, the system incorporating it and its application. Such analysis can help reduce the implementation costs of soft error mitigation methods.

3.5   Case Study: Evaluating the Impact of Aging on Silicon-Based Physically Unclonable Functions

This section presents a case study on the evaluation of aging impact on the quality metrics of PUF designs.

3.5.1   Evaluation Methodology

Modelling of aging effects is divided into two stages: the first is creating the aging-induced degradation model and second, integrating such a model into simulation tools. The R–D (reaction–diffusion) and the "lucky electron" models are used to model BTI and HCI as explained in the above sections. The use of these models requires extraction (i.e., through a curve fitting process) of some technology dependent parameters of their analytical model from burn-in experiments. The electrical parameters of the degraded devices are updated and integrated in their simulation models

Automated reliability tools like Cadence RelXpert [41], and Synopsis HSPICE MOSRA [42] apply similar processes with two stages of simulation. The first stage is called prestress, where the tool computes the degraded electrical of aged MOS transistors in the circuit based on circuit behaviour and on the built-in stress model, including BTI and HCI effects. The second stage is called post-stress, where the tool evaluates the impact of the degradation on the properties of the circuits (e.g. performance, leakage power…).

Simulation Tip: The same process as above can be used to evaluate aging impact on PUF designs. Access to the above aging tools is not necessarily required to carry first-order analysis, for example, the reader can compute the drift in the threshold voltages due to BTI, as given in Eq. (3.1) for a 32 nm technology. The new computed threshold voltage should then be used to update the spice models for the 32 nm technology library files. The aged models can then be used to build aged PUF design for evaluation. A conceptual diagram of the aging estimation methodology is shown in Fig. 3.5.

Fig. 3.5 Conceptual diagram
of an evaluation methodology
of Aging impact on
silicon-based PUFs

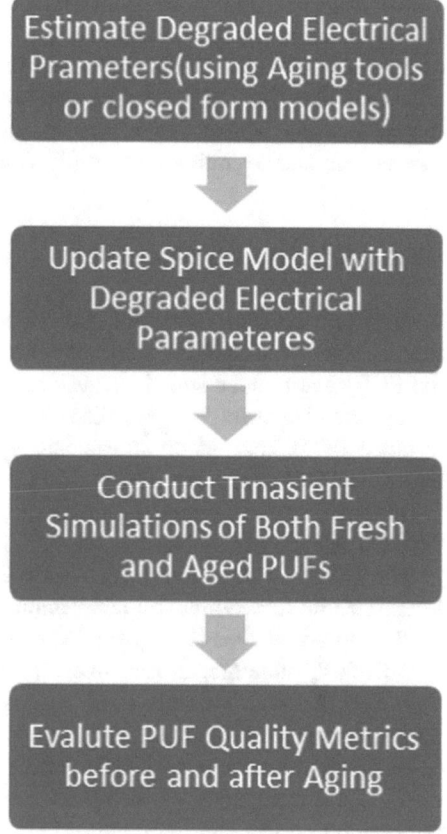

3.5.2   Experimental Setup

To illustrate how the above method can be applied, we consider three different PUF
designs as follows:

- Arbiter PUF which consists of 16 delay stages.
- TCO PUF implemented with a configuration of 8 × 4 array.
- SRAM PUF which consists of 4 byte × 64 rows.

The above designs are implemented with 65-nm technology node and the
BSIM4 (V4.5) transistor model. The HSPICE MOSRA tools are then used to
estimate the degradation in threshold voltage induced by 10 years of BTI-induced
aging. The activity factor (α) is assumed 20% (see Eq. 3.1); the actual value of this
factor is application dependent.

To model the impact of process variation on the physical parameters of CMOS
transistors (e.g. effective length, effective width, oxide thickness and threshold

voltage), we use Monte Carlo simulations with built-in fabrication standard statistical variation (3σ variations) in TSMC 65nm technology design kit. The source of threshold voltage variations is assumed to be caused by random dopant fluctuations (RDFs) [43]. For each of the above designs, a 100 PUF instances are modelled using Monte Carlo. Finally, the analysis has been done at a nominal temperature and operating voltage (25 C and 1.2 V).

3.5.3   Results and Discussion

To facilitate the analysis of the PUF metrics for all above designs, 32-bit lengths responses are considered. To estimate the impact of BTI aging on the reliability of the PUF circuits, identical challenges are applied on both fresh and aged designs, their respective responses are then compared, and in each case, the Hamming distance (HD) between the fresh and aged responses is computed. The same experiment is then repeated for the 100 instances of each PUF, and the bit error rate is computed as the average of all computed Hamming distances. The results are depicted in Fig. 3.6.

The TCO PUF seems to be resilient to BTI aging mechanism, this mainly due to its differential and symmetric architecture. In other words, the BTI stress patterns (in the form of challenge bits) applied to the Stochastic transistors in both of arrays are identical, therefore, in principles, the degradation in their electrical parameter should be identical. This effectively means comparing the current output of the two arrays should yield the same results before and after aging. However, there is a second-order effect that may cause error; in fact, BTI stress may affect the symmetry of the comparator, which makes it favour one input over the other. More information on this can be found in [44].

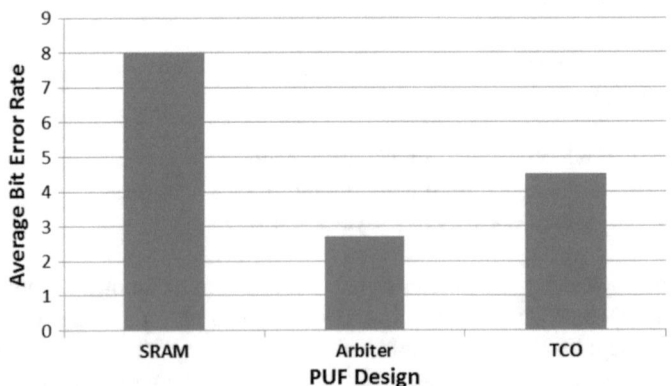

Fig. 3.6 The impact of BTI aging on the reliability of PUF designs

The Arbiter PUF also has a symmetric structure; this means its devices are likely to experience symmetrical stress patterns, assuming random challenges are applied throughout the lifetime.

In the case of SRAM PUF, the impact of aging is more pronounced.

Let us discuss this more closely. In the technology under consideration, the impact of PBTI is insignificant, so NBTI is the dominant aging mechanism. For a SRAM cell, seen in Fig. 3.7, only the pull-up transistors, P1 and P2 would suffer from NBTI. The pull-down transistors (N1 and N2) are only affected by PBTI, which causes negligible degradation in their electrical parameters. P1 and P2 transistors are part a cross-coupled inverter, this means only one would be under NBTI stress at any time. Such asymmetric stress conditions result in unbalanced threshold voltage degradation of these two transistors, for example, if P1 is placed under greater stress then its threshold voltage degradation will be bigger compared to P2, which will reduce its driving capability, consequently, node Q is less likely to be pulled up to be 1 once the cell is powered up.

If the initial start-up value of the above SRAM cell was 0 due to inherent process variation, such asymmetric aging can change this value to "1". This bit flipping effect reduces the capability of SRAM PUF to generate consistent responses based on its power-up values, which explains the results obtained in Fig. 3.6. Similar results are obtained in [45], where authors employed the static noise margin to quantify the impact of BTI on SRAM.

To estimate the impact of BTI aging on the uniqueness and uniformity metrics, the fresh and aged designs are evaluated separately. A summary of the result is shown in Table 3.1.

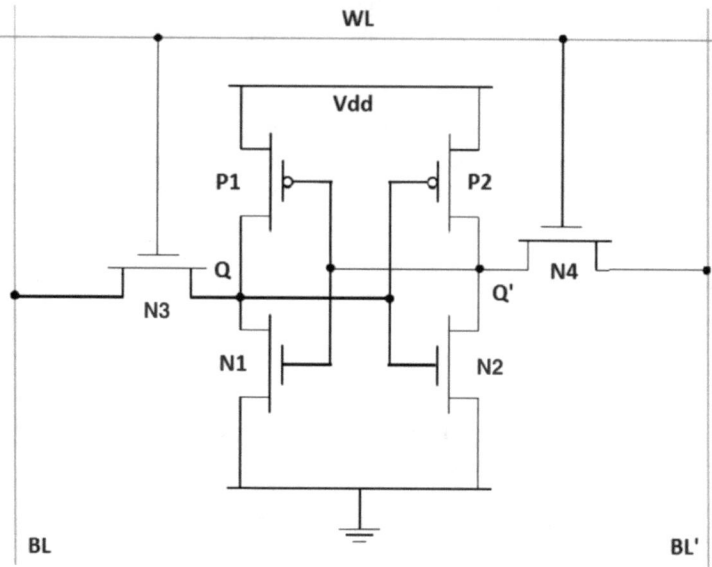

Fig. 3.7 A generic 6-transistor SRAM cell

Table 3.1 The impact of BTI aging on the uniqueness and uniformity of PUF Designs

PUF	Uniqueness (%)		Uniformity (%)	
	Fresh	Aged	Fresh	Aged
Arbiter	36.5	38.42	26.19	27.84
TCO	49.9	50.01	53.94	53.38
SRAM	50.01	49.94	49.78	48.95

The impact of BTI aging mechanism on the uniqueness metric of the PUF designs under consideration is negligible. Although errors induced by aging mechanism affect the response of the PUF design, such random flipping of bits is most likely to cancel out when computing the Hamming distance between these responses, therefore, the uniqueness metric remains almost the same.

The same rational can be applied to explain the insignificant degradation of the uniformity metric in the case studied here.

3.6   Conclusions and Learned Lessons

There are a number of temporal reliability challenges facing nano-scale CMOS devices, these can be categorised into two types, the first is marked by progressive degradation of their electrical parameters, which is typically termed CMOS aging, there are three considerable wear out mechanisms, namely, BTI, HCI and time-dependent dielectric breakdown. The second category is marked by transient errors, which can be caused by a variety of physical mechanisms including crosstalk, thermal noise, ground bounce and radiation hits. These reliability problems mainly affect the ability of the PUF to generate consistent responses (i.e. its reliability metric); other qualities of the PUF designs are less affected by these issues.

In order to devise a cost-effective mitigation measure, the impact of each of these physical phenomena needs to be carefully assessed, taking into consideration the type of the PUF, the nature of the system and the application used for. PUF-based key generation schemes are the most reliability-critical applications.

3.7   Problems

3.1   Which of the following PUF designs is most likely to be affected by radiation-induced soft errors?

(a)   A PUF embedded in a smart car application.
(b)   A PUF used for satellites identification.
(c)   A PUF used for building a controlled access security scheme.

3.2. A resistor has a value of R = 5000 Ω, and temperature T = 340 k. What is the magnitude of its thermal noise? Is this likely to cause an error in a SRAM PUF fabricated using 32 nm technology node? It is assumed that the threshold voltage of the MOS transistor in this technology is 0.53 V.

3.3. A PUF circuit, which has 3-bits challenge and 5-bits responses, is implemented on four chips. The challenge/response behaviour of this design is summarised in Table 3.2a below. The four chips are then subjected to accelerated aging process. The challenge response behaviour of the aged chips is summarised in Table 3.2b

  1. Compute the average bit error rate caused by the aging process.
  2. Table 3.3 shows the challenges response behaviour of the fresh and aged PUFs under different supply voltages. Analyze the given data and explain whether the aging process affects the robustness of the PUF design against fluctuations in the supply voltage.

Table 3.2 Challenge/response behaviour for different chips

Challenge	Device			
	Chip 1	Chip 2	Chip 3	Chip 4
(a) Fresh PUF				
000	11010110	01100001	10001001	11101101
001	00011100	00010000	11001010	01010101
010	10110100	11000111	01010100	11000011
011	10000101	10000101	10111011	01111111
100	11010011	11001010	01010100	11101001
101	10110101	11101000	11110101	11001111
110	11000101	01000011	01001000	11011000
111	11000101	01110101	01100001	01000101
(b) Aged PUF				
000	11010110	01100001	10001001	11101101
001	00011100	00010001	11001010	01010101
010	10110100	11000111	01010101	11000011
011	10000101	10000101	10111011	01111010
100	11010011	11001010	01010100	11101001
101	10110100	11101001	11110101	11001111
110	11000101	01000011	01001000	11011000
111	11000101	01110101	01100001	01000101

3.4. Which of the following PUFs are most likely to be affected by crosstalk noise?

   (a) A PUF integrated on a high-performance multicore processor chip.
   (b) A PUF integrated on a low power embedded system.

3.5. Which of the depicted PUFs in Fig. 3.8 would you recommend as aging resilient design? Explain your answer.

Table 3.3 Challenge/response behaviour at different supply voltages

Challenge	Device		
	Chip 1(V = 0.85 V)	Chip 1 (V = 1 V) (Nominal)	Chip 1 (V = 1.15 V)
(a) Fresh PUF			
000	11010000	11010001	11010001
001	11000010	11000000	11000010
010	10011000	10011001	10011001
011	11000000	11000001	11000001
100	10000111	10000110	10000111
101	11111110	11111111	11111111
110	10000111	10000110	10000110
111	11101001	11101000	11101001
(b) Aged PUF			
000	11010000	11010001	11010001
001	11000010	11000000	11000010
010	10011000	10011001	10011001
011	11000000	11000001	11000001
100	10000100	10000111	00000111
101	11111110	11111111	11111111
110	10000111	10000110	10000110
111	11101001	11101000	11101001

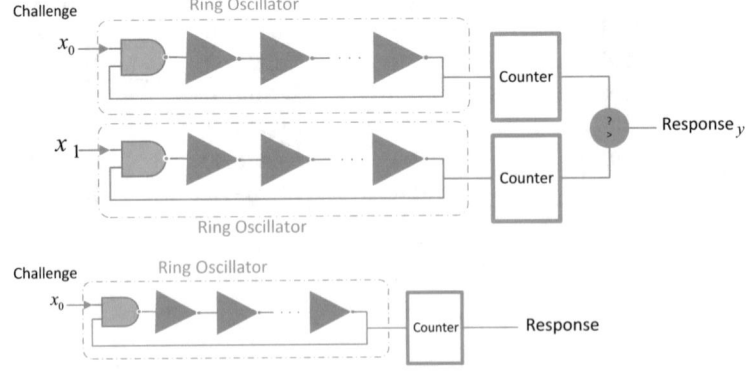

Fig. 3.8 Design choices for aging resilient PUF

References

1. V. Ferlet-Cavrois, L.W. Massengill, P. Gouker, Single event transients in digital CMOS—a review. IEEE Trans. Nucl. Sci. **60**, 1767–1790 (2013)
2. M. Agarwal, B.C. Paul, M. Zhang, S. Mitra, Circuit failure prediction and its application to transistor aging, in *VLSI Test Symposium, 2007. 25th IEEE* (2007), pp. 277–286
3. S. Bhardwaj, W. Wang, R. Vattikonda, Y. Cao, S. Vrudhula, Predictive modeling of the NBTI effect for reliable design, in *Custom Integrated Circuits Conference, 2006. CICC'06. IEEE* (2006), pp. 189–192
4. R. Vattikonda, W. Wenping, C. Yu, Modeling and minimization of PMOS NBTI effect for robust nanometer design, in *Design Automation Conference, 2006 43rd ACM/IEEE* (2006), pp. 1047–1052
5. D. Rossi, M. Omaña, C. Metra, A. Paccagnella, Impact of bias temperature instability on soft error susceptibility, in *IEEE Transaction on Very Large Scale Integration (VLSI) Systems*, vol 23 (2015), pp. 743–751
6. B.C. Paul, K. Kunhyuk, H. Kufluoglu, M.A. Alam, K. Roy, Impact of NBTI on the temporal performance degradation of digital circuits. IEEE Electron Devices Lett. **26**, 560–562 (2005)
7. H.I. Yang, C.T. Chuang, W. Hwang, Impacts of contact resistance and NBTI/PBTI on SRAM with high-? Metal-gate devices, in *IEEE International Workshop on Memory Technology, Design, and Testing, 2009. MTDT'09* (2009), pp. 27–30
8. H.K.M.A. Alam, D. Varghese, S. Mahapatra, A comprehensive model for pmos nbti degradation: recent progress. Microelectron. Reliab. **47**, 853–862 (2007)
9. M. Fukui, S. Nakai, H. Miki, S. Tsukiyama, A dependable power grid optimization algorithm considering NBTI timing degradation, in *IEEE 9th International New Circuits and Systems Conference (NEWCAS)* (2011), pp. 370–373
10. K. Joshi, S. Mukhopadhyay, N. Goel, S. Mahapatra, A consistent physical framework for N and P BTI in HKMG MOSFETs, in *Reliability Physics Symposium (IRPS), 2012 IEEE International* (2012), pp. 5A.3.1–5A.3.10
11. X. Li, J. Qin, J.B. Bernstein, Compact modeling of MOSFET wearout mechanisms for circuit-reliability simulation. IEEE Trans. Devices Mater. Reliab. **8**, 98–121 (2008)
12. F. Jianxin, S.S. Sapatnekar, Scalable methods for the analysis and optimization of gate oxide breakdown, in *2010 11th International Symposium on Quality Electronic Design (ISQED)* (2010), pp. 638–645
13. K. Weide-Zaage, Kludt, M. Ackermann, V. Hein, M. Erstling, Life time characterization for a highly robust metallization, in *2015 16th International Conference on Thermal, Mechanical and Multi-Physics Simulation and Experiments in Microelectronics and Microsystems* (2015), pp. 1–6
14. S. Moreau, D. Bouchu, Reliability of dual damascene TSV for high density integration: the electromigration issue, in *2013 IEEE International Reliability Physics Symposium (IRPS)* (2013), pp. CP.1.1–CP.1.5
15. S. Moreau, Y. Beilliard, P. Coudrain, D. Bouchu, R. Taïbi, L. D. Cioccio, Mass transport-induced failure in direct copper (Cu) bonding interconnects for 3-D integration, in *2014 IEEE International Reliability Physics Symposium* (2014), pp. 3E.2.1–3E.2.6
16. J. Ziegler, W. Lanford, The effect of sea level cosmic rays on electronic devices, in *1980 IEEE International Solid-State Circuits Conference. Digest of Technical Papers* (1980), pp. 70–71
17. T.C. May, M.H. Woods, Alpha-particle-induced soft errors in dynamic memories. IEEE Trans. Electron Devices **26**, 2–9 (1979)
18. D.C. Matthews, M.J. Dion, NSEU impact on commercial avionics, in *2009 IEEE International Reliability Physics Symposium* (2009), pp. 181–193
19. S. Uznanski, R.G. Alia, E. Blackmore, M. Brugger, R. Gaillard, J. Mekki et al., The effect of proton energy on SEU cross section of a 16 Mbit TFT PMOS SRAM with DRAM capacitors. IEEE Trans. Nucl. Sci. **61**, 3074–3079 (2014)

20. J.G. Rollins, W.A. Kolasinski, D.C. Marvin, R. Koga, Numerical simulation of SEU induced latch-up. IEEE Trans. Nucl. Sci. **33**, 1565–1570 (1986)
21. Y. Lin, M. Zwolinski, B. Halak, A low-cost radiation hardened flip-flop, in *2014 Design, Automation & Test in Europe Conference & Exhibition (DATE)* (2014), pp. 1–6
22. C. Slayman, Soft error trends and mitigation techniques in memory devices, in *2011 Proceedings—Annual Reliability and Maintainability Symposium* (2011), pp. 1–5
23. Y. Lin, M. Zwolinski, B. Halak, A low-cost, radiation-hardened method for pipeline protection in microprocessors. IEEE Trans. Very Large Scale Integr. VLSI Syst. **24**, 1688–1701 (2016)
24. M.A. Elgamel, K.S. Tharmalingam, M.A. Bayoumi, Crosstalk noise analysis in ultra deep submicrometer technologies, in *IEEE Computer Society Annual Symposium on VLSI* (2003), pp. 189–192
25. B.P. Wong, A. Mittal, Z. Gau, G. Starr, *Nano-CMOS circuits and physical design* (Wiley, Hoboken, New Jersey, 2005)
26. B. Halak, A. Yakovlev, Fault-tolerant techniques to minimize the impact of crosstalk on phase encoded communication channels. Comput. IEEE Trans. **57**, 505–519 (2008)
27. C. Duan, A. Tirumala, S.P. Khatri, Analysis and avoidance of cross-talk in on-chip buses, in *IEEE Conference on Hot Interconnects*, (2001), pp. 133–138
28. M.A. Elgamel, K.S. Tharmalingam, M.A. Bayoumi, Noise-constrained interconnect optimization for nanometer technologies. Int. Symp. Circ. Syst. **5**, 481–484 (2003)
29. M. Lampropoulos, B.M. Al-Hashimi, P. Rosinger, Minimization of crosstalk noise, delay and power using a modified bus invert technique. Des. Autom. Test Eur. Conf. Exhib. **2**, 1372–1373 (2004)
30. J. Nurmi, H. Tenhunen, J. Isoaho, A. Jantsch, *Interconnect Centric Design for Advanced SoC and NoC* (Kluwer Academic Publisher, Boston, 2004)
31. K.N. Patel, I.L. Markov, Error-correction and crosstalk avoidance in DSM busses, in *IEEE Transactions on Very Large Scale Integration (VLSI) Systems*, vol 12 (2004), pp. 1076–1080
32. D. Rossi, C. Metra, A.K. Nieuwland, A. Katoch, New ECC for crosstalk impact minimization. Des. Test Comput. IEEE **22**, 340–348 (2005)
33. D. Rossi, C. Metra, A.K. Nieuwland, A. Katoch, Exploiting ECC redundancy to minimize crosstalk impact. Des. Test Comput. IEEE **22**, 59–70 (2005)
34. A. Balasubramanian, A.L. Sternberg, B.L. Bhuva, L.W. Massengill, Crosstalk effects caused by single event hits in deep sub-micron CMOS technologies. Nucl. Sci. IEEE Trans. **53**, 3306–3311 (2006)
35. A.K. Palit, K.K. Duganapalli, W. Anheier, Crosstalk fault modeling in defective pair of interconnects. Integr. VLSI J. **41**, 27–37 (2008)
36. A. Kabbani, A.J. Al-Khalili, Estimation of ground bounce effects on CMOS circuits. IEEE Trans. Compon. Packag. Technol. **22**, 316–325 (1999)
37. S. Kim, C.J. Choi, D.K. Jeong, S.V. Kosonocky, S.B. Park, Reducing ground-bounce noise and stabilizing the data-retention voltage of power-gating structures. IEEE Trans. Electron Devices **55**, 197–205 (2008)
38. A. Antonopoulos, M. Bucher, K. Papathanasiou, N. Mavredakis, N. Makris, R.K. Sharma et al., CMOS small-signal and thermal noise modeling at high frequencies. IEEE Trans. Electron Devices **60**, 3726–3733 (2013)
39. Y.J. Lee, S.K. Lim, Co-optimization and analysis of signal, power, and thermal interconnects in 3-D ICs. IEEE Trans. Comput. Aided Des. Integr. Circ. Syst. **30**, 1635–1648 (2011)
40. B. Halak, Partial coding algorithm for area and energy efficient crosstalk avoidance codes implementation. IET Comput. Digit. Tech. **8**, 97–107 (2014)
41. C.D.S. Inc., Virtuoso relXpert reliability simulator user guide. Technical Report (2014)
42. S. Inc, *HSPICE User Guide: Basic Simulation and Analysis*. Technical Report (2013)
43. Y. Ye, F. Liu, M. Chen, S. Nassif, Y. Cao, Statistical modeling and simulation of threshold variation under random dopant fluctuations and line-edge roughness. IEEE Trans. Very Large Scale Integr. VLSI Syst. **19**, 987–996 (2011)

44. M.S. Mispan, B. Halak, M. Zwolinski, NBTI aging evaluation of PUF-based differential architectures, in *2016 IEEE 22nd International Symposium on On-Line Testing and Robust System Design (IOLTS)* (2016), pp. 103–108
45. M. Cortez, A. Dargar, S. Hamdioui, G.J. Schrijen, Modeling SRAM start-up behavior for Physical unclonable functions, in *2012 IEEE International Symposium on Defect and Fault Tolerance in VLSI and Nanotechnology Systems (DFT)* (2012), pp. 1–6

Reliability Enhancement Techniques for Physically Unclonable Functions

4

4.1 Introduction

The reliability of a physically unclonable device refers to its ability to generate the same response repeatedly given a specific challenge. This metric is an important factor that a designer must take into consideration when deciding on the suitability of a particular PUF architecture for a certain application. For example, cryptographic key generation schemes require the use of PUF devices which have 100% reliability (i.e. zero bit error rate), otherwise, encryption/decryption will not be possible. On the other hand, some classes of authentication applications can tolerate a certain level of bit error (i.e. the silicon device can be authenticated if its PUF circuit generates a response whose Hamming distance from the expected response is smaller than a predefined upper bound).

There are a number of mechanisms that cause the PUF responses to deviate from their expected values, these include variability in the environment conditions (i.e. temperature and power supply), electromagnetic interferences (i.e. ground bounce and crosstalk), radiation hits and aging of the underlying CMOS devices.

In order for a designer to ensure that the reliability of PUF responses adheres to the application requirements, they need first to evaluate the expected bit error rate at the output of the PUF and then apply the appropriate measures accordingly.

Reliability enhancement techniques of PUF designs can generally be classified into two categories, the first is based on the use of error correction codes and the second is based on pre-processing methods. The latter are applied at the post-fabrication stage before the chips are deployed to the field, the aim of such approaches is to reduce the likelihood of bit flipping, hence reducing or completely removing the need for expensive error correction schemes.

© Springer International Publishing AG, part of Springer Nature 2018
B. Halak, *Physically Unclonable Functions*,
https://doi.org/10.1007/978-3-319-76804-5_4

This chapter aims to:

1. Provide a comprehensive tutorial on the design and implementation principles of error correction schemes typically used for reliable PUF designs.
2. Explain in details the methods used for reliable PUF response construction schemes.
3. Discuss the state of the art pre-processing approaches applied at the post-fabrication stage to enhance the robustness of PUF designs.
4. Discuss the costs associated with the existing reliability enhancement techniques.

It is hoped that this chapter will give the reader the necessary theoretical background and skills to be able to implement appropriate reliability enhancements methods, which meet the specifications of their applications and the constraints of their design budgets.

4.2 Chapter Overview

The organisation of this chapter is as follows. Section 4.3 explains how to compute the expected bit error rate of a PUF, given a number of temporal noise sources. Section 4.4 presents a comprehensive tutorial on the design principles of error correction codes supported by examples and worked solutions. Section 4.5 outlines the design flow of reliable PUFs, including the details of each phase. Section 4.6 explains how the reliability of silicon-based PUFs can be enhanced using aging acceleration techniques. Section 4.7 describes the principles of the stable bit selection approach and gives two examples of how this technique can be implemented in practice to enhance the robustness of PUF designs. Section 4.8 explains the methods used for a reliable on-chip response construction, in particular, it discusses the use of secure sketch schemes, temporal majority voting and hardware redundancy. Section 4.9 analyses the costs associated with the above reliability enhancement approaches and discusses the implications of such costs on the use of PUF technology in resources constrained systems. Conclusions and lessons learned are presented in Sect. 4.10. Finally, a list of problems and exercises are provided in Sect. 4.11.

4.3 The Computation of the Bit Error Rate of PUFs

There are a number of physical sources of errors (e.g. aging, radiation ,etc.). If these contributions are assumed independent and additive, the overall error rate for a response bit (r_i) can be given as follows:

$$P_e(r_i) = \sum_{j=1}^{F} P_e^j(r_i), \tag{4.1}$$

where

F   is the total number of failure mechanisms.
P_e^j is the error rate caused by a failure mechanism j.

To compute the average error rate of all response bits, caused by a specific source of errors, we can use the formulae given below.

$$P_e^j = \frac{1}{k}\sum_{i=1}^{k} \frac{HD(R_i(n), R_i'(n))}{n} \times 100\%, \tag{4.2}$$

where

n The length of the PUF response (bits).
$HD(R_i(n), R_i'(n))$ The Hamming distance between response $Ri(n)$ from the chip
                              i at noise-free operating conditions and the same n-bit response
                              obtained at noisy conditions $R'i\,(n)$, respectively for the
                              challenge C.
K The number of sample chips.

We can also compute the average overall error rate caused by all sources of noise as follows:

$$P_e = \frac{1}{F}\sum_{j=1}^{F} P_e^j \tag{4.3}$$

It is possible in some cases to pre-compute the error rate using simulation models or test chips. Figure 4.1 shows how the bit error rate caused by temperature variations in different PUF architectures, namely:

- Arbiter PUF which consists of 16 delay stages.
- TCO PUF implemented with configuration of 8 × 4 array.
- SRAM PUF which consists of 4 byte × 64 rows.

The above designs are implemented using 65-nm technology node and the BSIM4 (V4.5) transistor model. For each of the above designs, a 100 PUF instances are modelled using Monte Carlo simulations. To facilitate the analysis of the PUF metrics for all of the above designs, 32-bit lengths responses are considered. The responses of each case are recorded at three different temperature (−40, 0 and 85 °C).

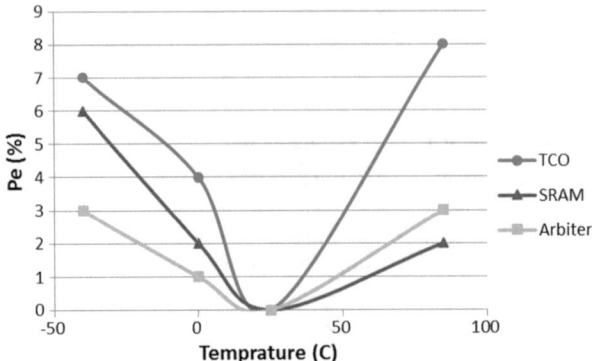

Fig. 4.1 The impact of temperature variations on the bit error rate of PUF architectures

The above experiment indicates that temperature variation can cause up to 8% errors. The arbiter design seems to be the most resilient to such variations under the above experimental conditions.

In order to get an accurate estimation of the overall expected bit error rate (as in Eq. 4.3), other sources of noise will also have to be included in the analysis, such as electromagnetic interference, CMOS aging and power supply fluctuations.

There is ongoing work to evaluate the impact of these mechanisms, for example, Maiti et al. in [1] presented a comprehensive study on how aging affects the reliability of delay-based physically unclonable functions, their estimate was in the region between 5 and 8% errors. Other papers also included similar estimations [2–4]. The impact of transient failure mechanisms has been comprehensively explored for CMOS-based designs, examples of which include the estimation of crosstalk-induced errors in [5] and the evaluation of radiation-induced soft errors [6, 7].

Designers of PUFs first need to identify the most probable cause of bit flips based on the architecture of the PUF and the type of system incorporating it, they can then devise appropriate methods to evaluate the expected bit error rate and the best reliability enhancement approach to adopt.

4.4 The Design Principles of Error Correction Codes

4.4.1 Preliminaries

A *block code* of *length n* consists of a set C of words (or strings) $a1a2 \ldots an$ of symbols ai taken from a finite alphabet A. If $|A| = r$, we call C an *r-ary code* (*binary* or *ternary* if $r = 2$ or $r = 3$). There are r^n such words, so $|C| \leq r^n$. The elements of C are called codewords.

A good way of correcting errors is to ensure that pairs of codewords do not resemble each other too closely. For example, if I make one or two typing errors in writing "Acknowledgement", you can probably still guess what I mean, since there are very few similar words in English, whereas even a single error in typing "bat" ("rat", "sat", "mat", "bar", etc.) could be totally confusing. In this context, it is useful to formally define the Hamming distance and the Hamming *weight*.

Definition 4.1

(a) **Hamming Distance** The Hamming distance $d(a, b)$ between two words $a = (a_i)$ and $b = (b_i)$ of length n is defined to be the number of positions, where they differ, that is, the number of $(i)s$ such that $a_i \neq b_i$.

(b) **Hamming Weight** Let 0 denotes the zero vector: 00 ... 0, The *Hamming weight* HW(a) of a word $a = a_1 ... a_n \in V$ is defined to be $d(a, 0)$, the number of symbols a$_i$! = 0 in a.

Example 4.1 The Hamming distance between "bat" and "bar" is 1.
The Hamming distance (d) has the following properties:

1. $d(a, b) \geq 0$ for all a and b, with equality if and only if a = b;
2. $d(a, b) = d(b, a)$ for all a and b;
3. $d(a, c) \leq d(a, b) + d(b, c)$ for all a, b and c (the triangle inequality).

Example 4.2 The Hamming distance between $a = 0100$ and $b = 1110$ is 2.
Error correction can be implemented using the nearest neighbour decoding. In this case, the receiver converts any received invalid codeword b to its nearest codeword $a \in C$ making an arbitrary choice, if there is more than one such codeword. Although this may or may not be the codeword that was transmitted, in many situations this simple rule minimises the error probability (i.e. the probability of incorrect decoding).

The minimum distance $d = d(C)$ of a code C is the minimum Hamming distance between any pair of distinct codewords, that is,

$$d = d(C) = min\{d(a,b)|a, b \in C, a \neq b\}$$

Example 4.3 What is the minimum distance of the code: $C = \{001, 011, 111\}$? And what is the Hamming weight of (011).

This can be computed by calculating the Hamming distances between each two words and then finding the minimum number, in this case, the answer will be one.

$$HW(011) = 2.$$

> **Theorem 4.1** *A code C can detect up to s errors if $d(C) \geq s + 1$. A code C can correct up to t errors if $d(C) \geq 2t + 1$.*

Example 4.4 Let C be the following repetition binary code $C = \{000, 111\}$, the minimum distance of this code is 3, this means it should be able to correct one error ($t = 1$). This can be clearly justified form observing the fact that a single error will transform the codeword (000) to another word (001), and using the nearest neighbouring principle, one can deduce that (001) is closer to (000) than to (111), so the receiver can correct this error and retrieve the originally submitted codeword.

The example above shows that ultimately a price has to be paid to correct errors, that is the sender will need to transmit more bits than otherwise required; those extra bits will be used for error correction.

We are now going to introduce a number of mathematical techniques useful for constructing error correction codes.

4.4.2 Groups

Let Z_n be a non-empty set, and let $*$ be a binary operation (such as addition or multiplication). We call $(Z_n, *)$ a group (G) if the following conditions are satisfied:

1. Closure: $\forall\ a, b \in G$ then $(a * b) \in Z_n$.
2. Associativity: $\forall\ a, b, c \in Z_n$ then $(a * b) * c = a * (b * c)$.
3. Identity: there exists $e \in Z_n$ such that $a * e = a = e * a$ for all $a \in Z_n$.
4. Inverses: for every $a \in Z_n$ there exists $ai \in G$ such that $a * ai = e = ai * a$.

Example 4.5 The number systems Z, Q, R, C and Z_n are groups under addition, with $* = +$, $e = 0$ and $ai = -a$.

A group G is said to be commutative (or abelian) if $(a * b) = (b * a)$ for all a, $b \in G$. (commutativity).

For example, the sets of non-zero elements in Q, R and C are all commutative groups under multiplication.

A group G is said to be cyclic if it has a generator element $g \in G$ such that every element $a \in G$ has the form $a = g^i$ (or ig in additive notation) for some integer i.

For example: Z_n is cyclic, since every element has the form $1 + 1+ \cdots + 1$. Z is also cyclic, since every element has the form $1 + 1+ \cdots + 1$. However, Q, R and C are not cyclic.

4.4.3   Rings and Fields

Definition 4.2 *Fields*
   A set F is a field if

1. It is a commutative group under addition.
2. Its nonzero elements form a commutative group under multiplication.
3. $a(b + c) = ab + ac$ for all $a, b, c \in F$ (distributivity).

For example, Q, R and C are all fields, but Z is not because the nonzero integer ± 2, ± 3 ,etc. do not have multiplicative inverses in Z (though they do in Q).

Definition 4.3 *Rings*
   A set is said to be a ring *(R)* if it has all the properties of a field except the existence of multiplicative inverses.

For example, Z is a ring. Another ring is Z_n, the latter is the set $\{0, 1 \dots n - 1\}$, where addition and multiplication defined as follows:

1. $+$: $a + b$ in $Z_n = (a + b)$ *mod n;*
2. $.$: $a . b$ in $Z_n = ab$ *mod n.*

It is useful at this stage to introduce the theorem below, which helps find fields:

Theorem 4.2 Z_n *is a field if and only if n is a prime number.*

For example, the integer 7 is a prime. In Z_7, we have $1 \times 1 = 2 \times 4 = 3 \times 5 = 6 \times 6 = 1$, so 1 and 6 are their own multiplicative inverses, while 2 and 4 are inverses of each other, as are 3 and 5. Thus, Z_7 is a field.

On the other hand, the integer 6 is not a prime, hence according to Theorem 4.2, Z_6 should not be a field. Let us check this: $6 = 2 \times 3$, so we have $2 \times 3 = 0$ mod 6,

Table 4.1 Addition and multiplication tables in the binary field Z_2

+	0	1
0	0	1
1	1	0

x	0	1
0	0	0
1	0	1

so the nonzero elements are not closed under multiplication, and cannot form a group. Thus, Z_6 is not a field.

A field containing only finitely many elements is called a *finite field*.

One of the most widely used field in computing applications is the binary field $Z_2 = \{1, 0\}$, Table 4.1 shows the addition and multiplication tables in this field.

We are now going to use the mathematical construction above in order to learn how to design linear codes, the most widely used form of error correction methods.

4.4.4 Fundamentals of Linear Codes

Consider the binary field $Z_n = \{0,1\}$, consider a set of words of length n ($a = \{a_1, a_2 \ldots a_n\}$, b, c …) constructed from elements from Z_2, for example, $a = \{1101\}$. These words can be considered as vectors.

Definition 4.4 *Vector Space*

 A set of q vectors defined over a commutative group Z_n will be considered as a vector space when its elements satisfy the following axioms:

1. $a \times (b * v) = (a \times b) * v$ *for all* $\alpha, b \in Z_n$ *and* $v \in V$.
2. $(a + b) * v = (a * v) \oplus (b * v)$ *for all* $\alpha, b \in Z_n$ *and* $v \in V$.
3. $a * (v \oplus u) = (a * v) \oplus (a * u)$ *for all* $\alpha \in Z_n$ *and* $u, v \in V$.
4. $1 * v = v$ *for all* $v \in V$.

where

$(\times, +)$ are addition and multiplication modulo n over the group, Z_n
$\oplus$ is a vector addition,
$*$ is a multiplication by a scalar (i.e. multiplication of a vector $v \in V$ by an element $a \in Z_n$).

For example, consider the vector spaces defined over the binary field $Z_2 = \{0, 1\}$:

$$V2 = \{0, 1\}$$
$$V4 = \{00, 01, 10, 11\}$$
$$V8 = \{000, 001, 010, 011, 100, 101, 110, 111\}$$

Definition 4.5 *Linear Codes*
   A subset $C \subseteq V(n, q)$ is a linear code, if $\alpha (a + b) = \alpha a + \alpha b$ for all a, $b \in C$ and for all $\alpha \in Z_n$. This means C should form a group under addition modulo n.

Definition 4.6 *Linearly Independent Vectors*
   A subset S of a space V defined over a field F is linearly independent if the equation: $\sum_{i=1}^{n} a_i s_i = 0$ necessarily implies that all $a_i = 0$.

For example, S1 = {100, 010, 001} *is linearly independent* because the equation $a_1 (100) + a_2 (010) + a_3 (001) = 0$ can only be satisfied if $a_1 = a_2 = a_3 = 0$.
   On the other hand, S2 = {100, 010, 110} is not linearly independent because the equation $a_1 (100) + a_2 (010) + a_3 (110) = 0$ can be satisfied when $a_1 = a_2 = a_3 = 0$ or when $a_1 = a_2 = a_3 = 1$.

4.4.5   How to Design Linear Codes

We are going to discuss two methods to construct linear codes; the first approach uses a set of linearly independent vectors as a basis for the code. The second technique uses a set of linear equations characterising the codewords.

4.4.5.1   Specifying Linear Codes Using Basis Vectors
The basis of the code C is a subset of the codewords of C, such that every codeword of C is a linear combination of these basis codewords. In other words, if a binary code of length n has a $(k \leq n)$ basis $(c_1, c_2 \ldots c_k)$, then any codewords (c_i) can be written as follows:

$$c_i = a_{i1}c_1 + a_{i2}c_2 + \cdots + a_{ik}c_k \quad where\{a_{i1}, a_{i2}\ldots a_{ik} \in Z_2\}$$

For example, assume $C = \{c1, c2, c3, c4\}$ has the basis $L = \{c1, c2\}$, this means that:

$$c_1 = a_{11}c_1 + a_{12}c_2$$
$$c_2 = a_{21}c_1 + a_{22}c_2$$
$$c_3 = a_{31}c_1 + a_{32}c_2$$
$$c_4 = a_{41}c_1 + a_{42}c_2, \quad where \{a_{11}, a_{12}, \ldots a_{42}\} \in Z_2$$

It is worth noting here that a binary code with a basis of k codewords will have 2^k codewords (i.e. it can encode k information bits).

For example, the code $C3 = \{0000, 0001, 1000, 1001\}$ has the basis: $L = \{0001, 1000\}$.

The dimension of code C with length n is defined as the number of basis vectors of this code, for example, $k(C3) = 2$, therefore, $n - k$ is the number of extra bits needed for error correction or detection (i.e. carry bits). The rate of linear code C with length n and a basis of k vector is defined as $\left(\frac{k}{n}\right)$.

> **Definition 4.7** *Generator Matrix*
> The generator matrix of a linear code is a binary matrix whose rows are the codewords belonging to the basis of this code. The dimension of this matrix is $(k \times n)$.

For example, the code $C3 = \{0000, 0001, 1000, 1001\}$ has the basis

$$L = \{0001, 1000\}.$$

Its generator matrix can be written as follows:

$$G = \begin{bmatrix} 0 & 0 & 0 & 1 \\ 1 & 0 & 0 & 0 \end{bmatrix}$$

It is worth noting here that the codewords of a code C with a generator matrix G ($k \times n$) can be calculated by multiplying all binary data words of size k by the generator matrix.

Example 4.6 Calculate the codewords of code $C3$ with the generator matrix:
$$G = \begin{bmatrix} 0 & 0 & 0 & 1 \\ 1 & 0 & 0 & 0 \end{bmatrix}.$$

Solution:

In this case, $n = 4$ and $k = 2$. Therefore, there are $2^k = 4$ codewords:

$$c_1 = [0 \quad 0] \cdot \begin{bmatrix} 0 & 0 & 0 & 1 \\ 1 & 0 & 0 & 0 \end{bmatrix} = [0 \quad 0 \quad 0 \quad 0]$$

$$c_2 = [0 \quad 1] \cdot \begin{bmatrix} 0 & 0 & 0 & 1 \\ 1 & 0 & 0 & 0 \end{bmatrix} = [1 \quad 0 \quad 0 \quad 0]$$

$$c_3 = [1 \quad 0] \cdot \begin{bmatrix} 0 & 0 & 0 & 1 \\ 1 & 0 & 0 & 0 \end{bmatrix} = [0 \quad 0 \quad 0 \quad 1]$$

$$c_4 = [1 \quad 1] \cdot \begin{bmatrix} 0 & 0 & 0 & 1 \\ 1 & 0 & 0 & 0 \end{bmatrix} = [1 \quad 0 \quad 0 \quad 1]$$

Definition 4.8 *Systematic Linear Codes*

A linear code is said to be **systematic** if it has a generator matrix $G = [I_k \quad P]$, where

I_k is the identity matrix with size $(k \times k)$

P is a matrix with the size $(k \times n - k)$.

We should note here that the generator matrix of a code C could be transformed into an equivalent systematic form using elementary row operation (which do not change the set of codewords) and column interchange.

For example, the generator matrix G below can be transformed into its systematic form (G_c) by interchanging the first and the seventh column

$$G = \begin{bmatrix} 0 & 0 & 0 & 0 & 1 & 1 & 1 \\ 1 & 1 & 0 & 0 & 1 & 0 & 0 \\ 1 & 0 & 1 & 0 & 0 & 1 & 0 \\ 1 & 0 & 0 & 1 & 1 & 1 & 0 \end{bmatrix}$$

$$G_C = \begin{bmatrix} 1 & 0 & 0 & 0 & 1 & 1 & 0 \\ 0 & 1 & 0 & 0 & 1 & 0 & 1 \\ 0 & 0 & 1 & 0 & 0 & 1 & 1 \\ 0 & 0 & 0 & 1 & 1 & 1 & 1 \end{bmatrix}$$

The main advantage of systematic codes is their ability to separate data and parity bits in the codewords. Take for example, the generator metric G_C in the above example.

Let $c_i = [c1, c2, c3, c4, c5, c6, c7]$ be a codeword generated from the data word $d_i = [d1, d2, d3, d4]$, we can write,

$$c_i = d_i * G_C$$

By replacing G from above, we can find

$$c1 = d1$$
$$c2 = d2$$
$$c3 = d3$$
$$c4 = d4$$
$$c5 = c1 + c2 + c4$$
$$c6 = c1 + c3 + c4$$
$$c7 = c2 + c3 + c4$$

The separation of data and parity bits can greatly reduce the complexity of hardware implementation of error correction codes.

4.4.5.2 Specifying Linear Codes Using Codewords Equations

Another way of specifying a linear code is to give a set of linear equations characterising the codewords.

$$h11 \times 1 + h12 \times 2 + \cdots + h1n \times n = 0$$
$$\vdots$$
$$hm1 \times 1 + hm2 \times 2 \cdots + hmn \times n = 0.$$

This means that all elements a $\in$ C should satisfy the equation:

$$aH^T = 0,$$

where

$$H = \begin{bmatrix} h_{11} & \cdots & h_{1n} \\ \vdots & \ddots & \vdots \\ h_{m1} & \cdots & h_{mn} \end{bmatrix}$$ is called the parity-check matrix size $(n - k, n)$,

H^T is the transpose of H,

0 denotes the $(k, n - k)$ zero matrix,

$a = \{a1, a2...an\}$ represent a codeword from C.

If there are m independent equations, then C has dimension $k = n - m$.

Let us now define the syndrome s of a codeword (a) from a code C, with a parity-check matrix H, as:

$$s = aH^T$$

It should be noted here that for all valid codewords $s = 0$ by definition, this property can be used to detect the presence of errors.

For example, if an erroneous codeword (a') is received then,

$$s' = aH^T \neq 0$$

The parity-check matrix H of a linear code with a generator matrix $G = \begin{bmatrix} I_k & P \end{bmatrix}$ can be computed as follows: $H = \begin{bmatrix} P^T & I_{n-k} \end{bmatrix}$.

4.4.6   Hamming Codes

Hamming codes are one of the most widely known examples of linear codes due to their relatively simple constructions.

For each integer $m \geq 2$, there is a Hamming code C defined over the binary field Z_2 and has the following parameters:

- Code length: $n = 2^m - 1$;
- Dimension: $k = 2^m - m - 1$;
- Minimum distance: $d = 3$.

The easiest way to describe a Hamming code is through its parity-check matrix.

First, we construct an $(m \times n)$ matrix whose columns are all non-zero binary m-tuples. For example, for a [7, 4] binary Hamming code, we take $m = 3$, so $n = 7$ and $k = 4$, we can obtain the parity-check matrix:

$$H = \begin{bmatrix} 1 & 0 & 1 & 0 & 1 & 0 & 1 \\ 0 & 0 & 1 & 0 & 0 & 1 & 1 \\ 0 & 0 & 0 & 1 & 1 & 1 & 1 \end{bmatrix}$$

Second, in order to obtain a parity-check matrix for a code in its systematic form, we move the appropriate columns to the end so that the matrix ends with the $(m \times m)$ identity matrix. The order of the other columns are irrelevant. The result is the parity-check matrix H for a Hamming $[n, k]$ code. In our example, we obtain:

$$Hc = \begin{bmatrix} 1 & 1 & 0 & 1 & 1 & 0 & 0 \\ 1 & 0 & 1 & 1 & 0 & 1 & 0 \\ 0 & 1 & 1 & 1 & 0 & 0 & 1 \end{bmatrix}$$

Third, based on H we can easily calculate the generator matrix (G) as follows:

$$G = [I_k \quad P],$$

where

$$H = [P^T \quad I_{n-k}]$$

$$H[7,4] = \begin{bmatrix} 1 & 1 & 0 & 1 & 1 & 0 & 0 \\ 1 & 0 & 1 & 1 & 0 & 1 & 0 \\ 0 & 1 & 1 & 1 & 0 & 0 & 1 \end{bmatrix} \text{ this means } P^T = \begin{bmatrix} 1 & 1 & 0 & 1 \\ 1 & 0 & 1 & 1 \\ 0 & 1 & 1 & 1 \end{bmatrix}$$

$$\text{So } G = \begin{bmatrix} 1 & 0 & 0 & 0 & 1 & 1 & 0 \\ 0 & 1 & 0 & 0 & 1 & 0 & 1 \\ 0 & 0 & 1 & 0 & 0 & 1 & 1 \\ 0 & 0 & 0 & 1 & 1 & 1 & 1 \end{bmatrix}$$

Now we have both H and G matrixes, we can proceed with encoding and decoding processes.

The codeword (c) of each data (d) word can be obtained by multiplication with the generator matrix

$$c = d.G$$

For example, to find the codeword for $d = (1001)$

$$c(1001) = d.G = \begin{bmatrix} 1 & 0 & 0 & 1 \end{bmatrix} \begin{bmatrix} 1 & 0 & 0 & 0 & 1 & 1 & 0 \\ 0 & 1 & 0 & 0 & 1 & 0 & 1 \\ 0 & 0 & 1 & 0 & 0 & 1 & 1 \\ 0 & 0 & 0 & 1 & 1 & 1 & 1 \end{bmatrix}$$

$$= \begin{bmatrix} 1 & 0 & 0 & 1 & 0 & 0 & 1 \end{bmatrix}$$

We notice that the first $(k = 4)$ bits of the codeword is the same as the data word, this is because we have used a systematic form of the G matrix, which makes decoding easier.

The decoding process has two stages: the first checks the validity of received codewords and makes correction, the second removes redundant bits (i.e. parity bits). This process assumes that the binary channel is symmetric, which means the probabilities of a bit "0" flip equal to the probability of bit "1" flip. The maximum likelihood decoding strategy is adopted, which means an erroneous codeword should be restored to a valid codeword that is closest to it in terms of Hamming distance. In case of a one error correcting Hamming code, only one error is assumed per codeword. The decoding process can be summarised in the following four steps:

1. Compute the syndrome for the received codeword (c) as follows: $s = cH^T$.
2. If ($s = 0$), then c is a valid codeword (by definition).
3. Otherwise, determine the position j of the column of H that is the transpose of the syndrome.
4. Change the jth bit in the received word, and output the resulting code.

As long as there is at most one-bit error in the received vector, it will be possible to obtain the original codeword that was sent.

Example 4.7 The [15, 11] binary systematic Hamming code has parity-check matrix

$$\begin{pmatrix} 0 & 0 & 0 & 0 & 1 & 1 & 1 & 1 & 1 & 1 & 1 & 1 & 0 & 0 & 0 \\ 1 & 1 & 1 & 0 & 0 & 0 & 0 & 1 & 1 & 1 & 1 & 0 & 1 & 0 & 0 \\ 0 & 1 & 1 & 1 & 0 & 1 & 1 & 0 & 0 & 1 & 1 & 0 & 0 & 1 & 0 \\ 1 & 0 & 1 & 1 & 1 & 0 & 1 & 0 & 1 & 0 & 1 & 0 & 0 & 0 & 1 \end{pmatrix}$$

Assume the received vector is

$$c = (0 \quad 0 \quad 0 \quad 0 \quad 1 \quad 0 \quad 0 \quad 0 \quad 0 \quad 0 \quad 1 \quad 1 \quad 0 \quad 0 \quad 1).$$

Check if c is a valid codeword. If not, compute the error vector.
Solution:

1. The syndrome is calculated to be $s = cH^T = (1 \quad 1 \quad 1 \quad 1)$.
2. Notice that s is the transpose of the 11th column of H, so we change the 11th bit of c to get the decoded word as $c = (1,0,0,1,0,0,0,0,1,0,0,1,0,0,1)$.
3. Since the first 11 bits give the information (as this code is systematic), the original message was $d = (1,0,0,1,0,0,0,0,1,0,0)$.

4.4.7 Cyclic Codes

Cyclic codes are a special type of codes, which allow for efficient hardware implementation; therefore, they can be particularly useful for resource-constrained systems such as IoT devices. In this section, we will learn their design principles; we will also explain how they can be realised in hardware.

4.4.7.1 Fundamentals

Definition 4.9 *Cyclic Linear Codes*
  A linear code C is cyclic if any cyclic shift of a codeword is also a codeword. In other words, whenever $a_0, \dots a_{n-1} \in C$, then also $a_{n-1} a_0 \dots a_{n-2} \in C$.

For example, the code $C1 = \{000, 101, 011, 110\}$ is cyclic, whereas the code $C2 = \{0000, 1001, 0110, 1111\}$ is not cyclic.

Let us take another example, consider the code $C3$ that has the following generator matrix:

$$G = \begin{pmatrix} 1 & 0 & 1 & 1 & 1 & 0 & 0 \\ 0 & 1 & 0 & 1 & 1 & 1 & 0 \\ 0 & 0 & 1 & 0 & 1 & 1 & 1 \end{pmatrix}$$

This code has the following codewords:

$$c_1 = 1011100 \quad c_2 = 0101110 \quad c_3 = 0010111$$
$$c_4 = 1110010 \quad c_5 = 1001011$$
$$c_6 = 0111001 \quad c_7 = 1100101$$

A closer inspection of the codewords, reveals the fact that this is a cyclic code because a right cyclic of each codeword would produce another valid codeword, for example, shifting c_1 produces c_1 and so on.

We are now going to introduce the mathematical tools needed to construct cyclic codes.

4.4.7.2 Rings of Polynomials

We have so far learned how to carry out modular addition and multiplication over the binary group, i.e. Z_2, we will now discuss how to carry out modular operation over a polynomial ring. In this case, we are only interested in polynomial rings defined modulo $(X^n + 1)$, over the binary field Z_2, but the lessons learned can be applied to all polynomial rings.

Theorem 4.3 *The set of polynomials over a field F of degree less than n form a ring with respect to addition and multiplication modulo $(X^n + 1)$. This ring will be denoted $F_q[X]/(X^n + 1)$.*

For example: the set of elements $Z_2[x]/(X^2 + 1)$ are $\{0, 1, X, X + 1\}$ in order to find the result of multiplying two elements (x) and $(x + 1)$, we first multiply these two polynomials and get $(x^2 + x)$, then we find the result by computing the remainder of dividing the resulting term over $x^2 + 1$, which is in this case $(1 + x)$. It worth noting that addition and multiplication of coefficients are performed using the

Table 4.2 Multiplication and addition tables in the ring $Z_2[x]/(X^2 + 1)$

*	0	1	x	$1 + x$
0	0	0	0	0
1	0	1	x	1 + x
x	0	x	1	1 + x
1 + x	0	1 + x	1 + x	0

+	0	1	x	$1 + x$
0	0	1	x	1 + x
1	1	0	1 + x	x
x	x	1 + x	0	1
1 + x	1 + x	x	1	0

same rules in the binary field Z_2 (e.g. $1 + 1 = 0$, $X + X = 0$, $X^2 + X^2 = 0$, etc.). The multiplication and addition tables in the ring above are included in Table 4.2.

4.4.7.3 Designing Cyclic Codes: The Generator Polynomial

Cyclic codes of length n use the multiplicative and additive structure of the ring $R_n = Z[x]/(X^n + 1)$, where codewords of length n such as $(a_{n-1} \, a_{n-2}...a_0)$ are associated with polynomials of degree $n - 1$, such as $a_{n-1} x^{n-1} +... a_2 x^2 + a_1 x + a_0$. For example, the codeword $m = 1011$ would be associated with polynomial $m(x) = x^3 + x + 1$.

The use of ring R_n makes it easier to perform multiplication, as we will see from the theorem below:

Theorem 4.4 *Multiplication of a cyclic codeword a(x) by x in R_n corresponds to a single cyclic shift.*

Proof

$$x.a(x) = x.\left(a_0 + a_1 x +...a_{n-1}x^{n-1}\right)$$
$$x.a(x) = a_0 x + a_1 x^2 +...+ a_{n-2}x^{n-1} + a_{n-1}x^n$$

Since $x^n = 1 \pmod{x^n + 1}$, this means

$$x.a(x) = a_0 x + a_1 x^2 +...+ a_{n-2}x^{n-1} + a_{n-1}$$

which can be rewritten as

$$x.a(x) = a_{n-1} + a_0 x + a_1 x^2 +...+ a_{n-2}x^{n-1}$$

To construct a cyclic code, we will need to define a condition that can be only satisfied by valid codewords. To do this, we can use the concept of the generator polynomial $g(x)$ as defined formally below.

Theorem 4.5 *Let C be an (n, k) linear cyclic code over the ring $R_n = Z[x]/(X^n + 1)$*

1. *There exists a monic polynomial* g(x), called the generator polynomial, such that n-tuple **c(x) is a codeword if and only if g(x) is a divisor of c(x)**.*
2. *The generator polynomial is **unique**.*
3. *The degree of the generator polynomial is **n − k**.*
4. *The generator polynomial is a divisor of **$x^n + 1$**.*
5. *for every data word **m(x)**, and its corresponding codeword **c(x)** the following relation holds: **c(x) = m(x)g(x)**.*

*A Monic Polynomial is A polynomial $(a_0 x + a_1 x^2 + \ldots + a_{n-2} x^{n-1} + x^n)$ in which the coefficient of the highest order term is 1.

The fourth claim of the previous theorem gives a recipe to get all cyclic codes of a given length n.

Indeed, all we need to do is to find all factors of $x^n + 1$, which can consequently be used as generator polynomials.

Example 4.8 Find all binary cyclic codes of length 3.

Solution:
According to Theorem 4.5 (claim 4), the generator polynomial should be a divisor of $x^3 + 1$:

First, we need to find all divisors of $x^3 + 1$

$$x^3 + 1 = (x^3 + 1).1$$
$$x^3 + 1 = (x + 1)(x^2 + x + 1)$$

This shows that we have four divisors $(x^3 + 1)$, 1, $(x + 1)$ and $(x^2 + x + 1)$, each of these can be a generator of a cyclic code. Therefore, we have in total four cyclic codes.

Second, in order to obtain the codewords for each of these codes, we multiply the corresponding generator polynomial by the elements of $R_3 = \{0, 1, x, 1 + x, x^2, 1 + x^2, x + x^2, 1 + x + x^2\}$.

- **For g(x) = 1**, we obtain the following codewords:
  $C(x) = \{0, 1, x, 1 + x, x^2, 1 + x^2, x + x^2, 1 + x + x^2\}$, which corresponds to
  $C = \{000,001,010,011,100,101,110,111\}$

This results in a code with $k = 3$ and $n = 3$. It is called a *no parity* cyclic code

- **For g(x) = x^3 + 1**, we obtain the following codewords:
  $C = \{0\} = \{000\}$

This results in a code with $k = 0$ and $n = 3$. It is called a *no information* cyclic Code

- **For g(x) = 1 + x**, we obtain the following codewords:
  $C(x) = \{0, 1 + x, x + x^2, 1 + x^2\}$, which corresponds to $C = \{000, 110, 011, 101\}$

This results in a code with $k = 2$ and $k = 3$. It is called a one parity cyclic code, capable of detecting one error,

- **For g(x) = x^2 + x + 1**, we obtain the following codewords:
  $C(x) = \{0, 1 + x + x^2\}$, which corresponds to $C = \{000, 111\}$

This results in a code with $k = 1$ and $n = 3$. It is called a repetition code capable of correcting one error.

Now that we have learned how to compute the generators polynomials of cyclic codes, we can discuss how to construct their associated generator matrixes. In fact, the latter can be constructed based on Theorem 4.6 below:

Theorem 4.6 *Suppose C is a cyclic code of codewords of length n with the generator polynomial*

$$g(x) = g_0 + g_1 x + \ldots + g_r x^r,$$

where $r = n - k$.
Then, dim $(C) = n - r$ and a generator matrix G for C is

$$G = \begin{pmatrix} g_0 & g_1 & g_2 & \cdots & g_r & 0 & 0 & 0 & \cdots & 0 \\ 0 & g_0 & g_1 & g_2 & \cdots & g_r & 0 & 0 & \cdots & 0 \\ 0 & 0 & g_0 & g_1 & g_2 & \cdots & g_r & 0 & \cdots & 0 \\ \cdots & \cdots & & & & & & & & \cdots \\ 0 & 0 & \cdots & 0 & 0 & \cdots & 0 & g_0 & \cdots & g_r \end{pmatrix}$$

G is a cyclic matrix, wherein each row is obtained by shifting the previous row to the right.

Proof First, we can easily notice that all rows of G are linearly independent.

Second, the $n-r$ rows of G represent codewords: $g(x), xg(x), x^2g(x), ..., x^{n-r-1}g(x)$.

Finally, It remains to show that every codeword in C can be expressed as a linear combination of vectors from $g(x), xg(x), x^2g(x), ..., x^{n-r-1}g(x)$.

Indeed, if $a(x) \in C$, then

$$a(x) = q(x)g(x).$$

Since $deg\ a(x) < n$, we have $deg\ q(x) < n - r$.

Hence,

$$q(x)g(x) = \left(q_0 + q_1x + ... + q_{n-r-1}x^{n-r-1}\right)g(x)$$
$$= q_0g(x) + q_1xg(x) + ... + q_{n-r-1}x^{n-r-1}g(x).$$

For example, consider the code $C = \{000, 111\}$ with $g(x) = x^2 + x + 1$, in this case, $n = 3$, $k = 1$ and the generator polynomial will have the dimensions $[3, 1]$ and can be constructed as follows:

$$G = \begin{bmatrix} 1 & 1 & 1 \end{bmatrix}.$$

Another example is the code $C = \{000, 110, 011, 101\}$ with $g(x) = x + 1$, in this case, $n = 3$, $k = 2$ and the generator polynomial will have the dimensions $[3, 2]$ and can be constructed as follows:

$$G = \begin{bmatrix} 1 & 1 & 0 \\ 0 & 1 & 1 \end{bmatrix}$$

4.4.7.4 Designing Cyclic Codes: The Parity Check Polynomial

Definition 4.10 *Parity-Check Matrix*

Let C be a cyclic $[n, k]$-code with the generator polynomial $g(x)$ (of degree $n - k$). By theorem 4.5, $g(x)$ is a factor of $x^n + 1$. Hence, $x^n + 1 = g(x)h(x)$ for some $h(x)$ of degree k. $h(x)$ is called the parity check polynomial of C. It is given as follows:

$$h(x) = h_0 + h_1x ... + h_kx^k$$

The parity-check matrix for this code $H(n - k, n)$ can be derived as follows:

$$H = \begin{pmatrix} h_k & h_{k-1} & h_{k-2} & \cdots & h_0 & 0 & 0 & 0 & \cdots & 0 \\ 0 & h_k & h_{k-1} & h_{k-2} & \cdots & h_0 & 0 & 0 & \cdots & 0 \\ 0 & 0 & h_k & h_{k-1} & h_{k-2} & \cdots & h_0 & 0 & \cdots & 0 \\ \cdots & \cdots & & & & & & & & \cdots \\ 0 & 0 & \cdots & 0 & 0 & \cdots & 0 & h_k & \cdots & h_0 \end{pmatrix}$$

Example 4.9 Find the parity check polynomial and its corresponding matrix for the code: $C = \{000, 110, 011, 101\}$ with $g(x) = x + 1$ defined over $R_3 = Z[x]/(X^3 + 1)$.

From Definition 4.10, the parity check polynomial is given as follows:

$$X^3 + 1 = g(x)h(x)$$

This means:

$$h(x) = \frac{x^3 + 1}{x + 1} = x^2 + x + 1$$

based on this, the parity-check matrix with the dimension $H(n - k, n) = H(1, 3)$ can be constructed as follows:

$$H = \begin{bmatrix} 1 & 1 & 1 \end{bmatrix}$$

The parity check polynomial can be used to check the validity of received codewords using the theorem 4.7 given below:

Theorem 4.7 *If an n-tuple c(x) is a valid codeword in a cyclic code C(n, k) with generator polynomial g(x) and a check polynomial h(x); the following relation holds: s(x) = c(x)h(x) ≡ 0 (modulo x^n + 1) where s(x) is the syndrome polynomial.*

Example 4.10 Consider code: $C = \{000, 110, 011, 101\}$ with $h(x) = x^3 + x + 1$ defined over $R_3 = Z[x]/(X^3 + 1)$, check that all codewords satisfy the theorem 4.7.

Solution:

The codewords in C can be written in their polynomial forms as follows:

$$C = \{0, 1 + x, x + x^2, 1 + x^2\}.$$

Multiplying each codeword with $h(x)$ gives the following (remember the multiplication is done modulo $(X^3 + 1)$).

$$(0)(x^2 + x + 1) = 0 \text{ modulo } (x^3 + 1)$$

$$(1 + x)(x^2 + x + 1) = x^2 + x + 1 + x^3 + x^2 + x = x^3 + 1 = 0 \text{ modulo } (x^3 + 1)$$

$$(1 + x^2)(x^2 + x + 1) = x^2 + x + 1 + x^4 + x^3 + x^2$$

$$= x^4 + x + x^3 + 1$$

$$= x(x^3 + 1) + (x^3 + 1) = 0 \text{ (Note that the term } (x^3 + 1)$$

equates to 0 in modulo $(x^3 + 1)$)

$$(x + x^2)(x^2 + x + 1) = x^3 + x^2 + x + x^4 + x^3 + x^2$$

$$= x^4 + x$$

$$= x(x^3 + 1) = 0 \text{ modulo } (x^3 + 1)$$

Executive Summary A Cyclic code (n, k) defined over a ring $Rn = Z[X]/(X^n + 1)$ will have a generator polynomial g(x) of degree $(n - k)$, such that all codewords can be derived using:

$$c(x) = d(x)g(x), \quad \text{where } d(x) \in R_n$$

This code will also have parity check polynomial (h(x)) such that:

$$g(x)h(x) = 1 + x^n \equiv 0(mod\ 1 + x^n)$$

All valid codewords should satisfy:

$$c(x)h(x) \equiv 0(mod\ 1 + x^n)$$

4.4.7.5   Hardware Implementation of Cyclic Encoders

The special structure of cyclic codes allows for efficient hardware implementation of encoding and decoding circuits.

To design a cyclic encoding circuit, the **last-in-first-out** shift register can be used as shown in Fig. 4.2, where g_0, g_1 ... g_{n-k} are the coefficients of the corresponding generator polynomial.

Encoding can be performed by serially inputting the k-bit message $m(x)$ followed by $n - k$ zeros and the output is the n-bit message $c(x)$.

Example 4.11 Construct the encoder for a cyclic code with the generator polynomial $g(x) = x^3 + x + 1$ and encode the message $m(x) = 1 + x + x^3$.

In this case, the coefficients of the generator polynomials are $\{g_0, g_1, g_3, g_3\} = \{1,1,0,1\}$, replacing this in the structure presented in Fig. 4.2 produces the encoder below.

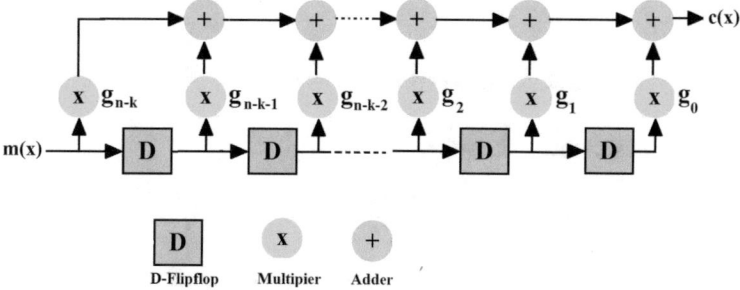

Fig. 4.2 Generic hardware implementation of cyclic encoder

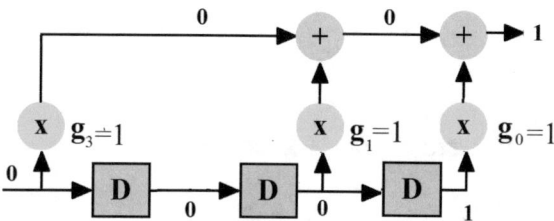

To encode the message $m = \{1101\}$, first the register are reset, then the message is inputted serially followed by $n - k$ zeros (in this case 3 zeros). The codeword is sampled at the output such that its most significant bit comes first. Let us study this closely, the applied data bit is shown in red.

Stage 1:

The first bit of the data word (highlighted in red) is applied which generates the coefficient of the largest term in the codewords.

Dataword: $m(x) = 1 + x + x^3 = 1101$

Codeword: $c(x) = x^6$

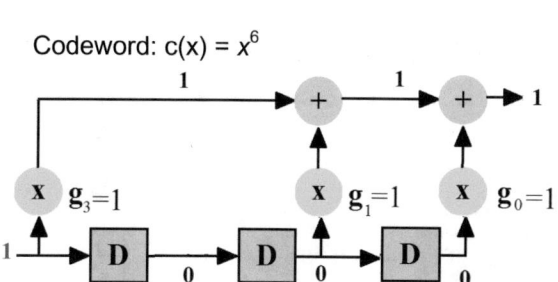

Stage 2:

Dataword: $m(x)= 1+ x +x^3 =1101$

Codeword: $c(x) = 0x^5 + x^6$

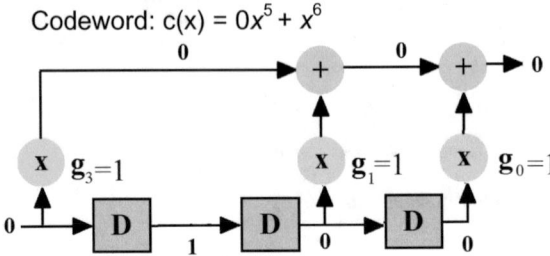

Stage 3:

Dataword: $m(x)= 1+ x +x^3 =1101$

Codeword: $c(x) = 0x^4 + 0x^5 + x^6$

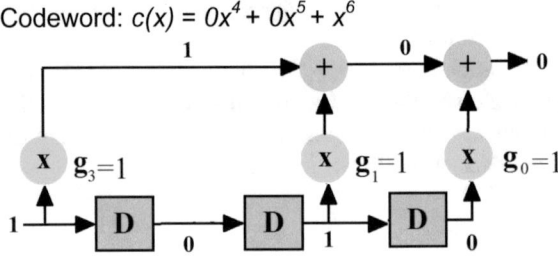

Stage 4:

Dataword: $m(x)= 1+ x +x^3 =1101$

Codeword: $c(x) = 0x^3 + 0x^4 + 0x^5 + x^6$

Stage 5: we now start feeding the input with zeros.

Dataword: m(x)= 1+ x +x^3 =1101

Codeword: c(x) = x^2 + $0x^3$ + $0x^4$ + $0x^5$ + x^6

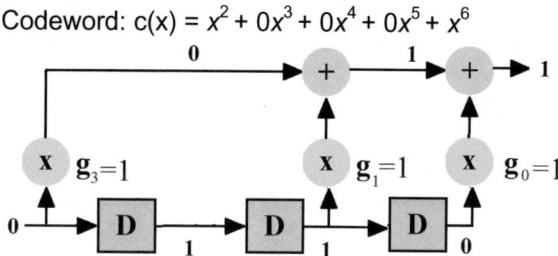

Stage 6:

Dataword: m(x)= 1+ x +x^3 =1101

Codeword: $c(x)$ = $0x$ + x^2 + $0x^3$ + $0x^4$ + $0x^5$ + x^6

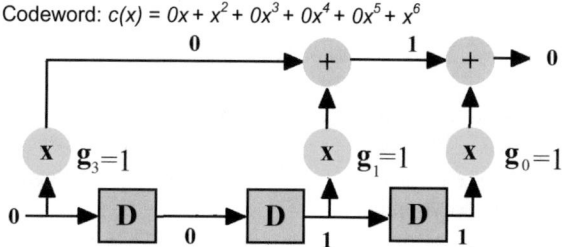

Stage 7: this is the final stage.

Dataword: m(x)= 1+ x +x^3 =1101

Codeword: c(x) = $1+0x$ + x^2 + $0x^3$ + $0x^4$ + $0x^5$ + x^6 = 1010001

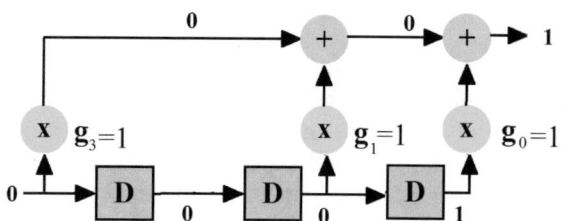

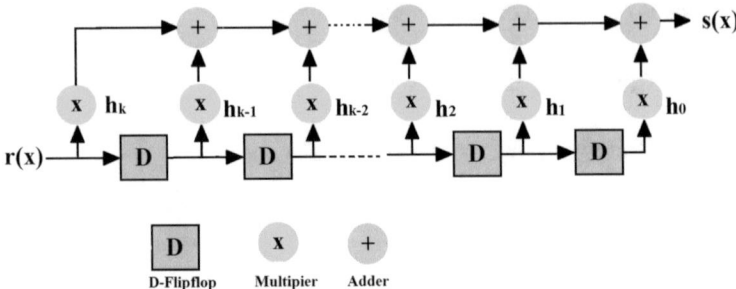

Fig. 4.3 Generic hardware implementation of a syndrome computing circuit for cyclic codes

4.4.7.6   Hardware Implementation of Cyclic Decoders

In order to verify whether or not a received codeword is valid, we can compute its syndrome based on Theorem 4.7.

Let $r(x)$ be the polynomial of a received codeword, the syndrome polynomial for this codeword is:

$$s(x) = r(x)h(x)$$

If $r(x)$ is a valid codeword, than $s(x) = 0$, otherwise $r(x) \neq 0$.

To design a syndrome computing circuit, the **last-in-first-out** shift register can be used as shown in Fig. 4.3, where $h_0, h_1 \dots h_k$ are the coefficients of the corresponding parity check polynomial.

The syndrome consists of the coefficients of $x^k \dots x^{n-1}$ in the resulting $s(x) = r(x) h(x)$, these are generated after $r_{n-1} \dots r_{n-k}$ have been shifted into the register.

It should finally be noted that for error correcting cyclic codes, additional circuitry may be required. The hardware requirements for these extra circuits can be minimised by using systematic codes wherein data and parity bits are separate (please see Sect. 4.4.5 for more details on the structure of systematic codes).

4.4.8   Golay Code

This is an important example of a cyclic code that can correct up to three errors. NASA used it in space missions in the early 80s and the new American government standards for automatic link establishment in high-frequency radio systems. From the PUF application viewpoint, this code offers the possibility of constructing a multi-bit correcting error scheme at relatively low cost using the shift register-based implementations explained in Sect. 4.4.7. Either of the following two polynomials can generate this code:

$$g_1(x) = 1 + X^2 + X^4 + X^5 + X^6 + X^{10} + X^{11}$$
$$g_2(x) = 1 + X + X^5 + X^6 + X^7 + X^9 + X^{11}$$

4.4.9   BCH Codes

These are a special class of cyclic codes with the ability to correct more than one error. They also have effective decoding procedures. They are named after Bose, Ray-Chaudhuri and Hoquenghem.

Hardware implementation of the encoding and syndrome computing blocks of the BCH codes are the same as that of cyclic codes seen previously. To understand the mathematical constructions of these codes, we will need to introduce the principles of Galois fields and primitive polynomials.

4.4.9.1   Galois Fields

> **Definition 4.11** *Irreducible Polynomials*
> A polynomial $p(x) \in Z_p[X]$ of degree $n \geq 1$ is irreducible over Z_p if $p(x)$ cannot be expressed as the product $p(x)\ r(x)$ of two polynomials $q(x)$ and $r(x)$ in $Z_2[X]$, where the degree of both $q(x)$ and $r(x)$ is greater than or equal to *1* but less than *n*.

It is worth noting here that an irreducible polynomial of degree $(n > 1)$ over a field F does not have any roots in F.

Example 4.12 Which of the following polynomials is irreducible in $Z_2[x]$?

$$f_1(X) = X^3 + X$$
$$f_2(X) = X^3 + 1$$
$$f_3(X) = X^3 + X + 1$$

Solution:

$f_1(X)$ is reducible as it has the root (0) in the binary field Z_2
$f_2(X)$ is reducible as it has the root (1) in the binary field Z_2
$f_3(X)$ is irreducible as it does not have roots in the binary field Z_2 and cannot be written as a product of two polynomial in $Z_2[x]$ of lesser degrees.

Definition 4.12 *Galois Field*

For any prime number p and positive integer n, there is a finite field Fq with $q = p^n$ elements, it is called the Galois field of order p^n and is denoted by $GF\ (q)$. The elements of Fq are the polynomials $(a_{n-1}X^{n-1} + \cdots + a_1X + a_0)$ of degree less than n in $Z_p[x]$ with the addition, subtraction and multiplication, modulo of an irreducible polynomial of degree n.

The multiplicative group F^*_q of the nonzero elements of F_q is cyclic (i.e. it has a generator), this generator is the root of the irreducible polynomial.

Example 4.13 Construct a Galois field $GF(2^2)$ and find its elements:

Solution:

$$p = 2, n = 2$$

$Z_2[X]$ is the set of polynomials with coefficients of Z_2.

We can find that $r(x) = 1 + x + x^2$ is an irreducible polynomial of degree $(n = 2)$ over Z_2.

The elements of $GF(2^2)$ can be found as the remainders of the division of $Z_P[X]$ polynomials over $r(x)$.

$$GF(2^2) = \{0, 1, x, x+1\}$$

Definition 4.13 *Primitive Element* (α)

A primitive element of a finite field $GF(q)$ is a generator of the multiplicative group of the field. This means that all the nonzero elements of $GF(q)$ can be expressed as α^i, wherein i is an integer.

Definition 4.14 *Primitive Polynomials*

An irreducible polynomial $f(X)$ of degree n in $Z_p[x]$ is primitive if its root (α) is a primitive element of the finite field F_q (i.e. it generates the multiplicative group of the resulting field F^*_q, where $(q = p^n)$).

Example 4.14 Consider the field $GF(2^2) = Z_2[x]/(1 + x + x^2) = \{0, 1, X, X + 1\}$.

Given that $r(x) = 1 + x + x^2$ is an irreducible polynomial of degree $(n = 2)$ over Z_2, verify whether or not it is primitive.

Solution:
Let us assume that α is a root of r(x).
This means

$$\alpha^2 + \alpha + 1 = 0$$
$$\alpha^2 = \alpha + 1$$

we can use the above equation to produce all the elements of $GF(4) = \{0, 1, x, x+1\}$, by computing the powers of α as follows:

$$\alpha^0 = 1$$
$$\alpha = \alpha$$
$$\alpha^2 = \alpha + 1$$
$$\alpha^3 = \alpha.\alpha^2 = \alpha(1 + \alpha) = \alpha^2 + \alpha = 1 + \alpha + \alpha = 1$$

This shows that every nonzero element of $GF(4)$ is a power of α, therefore r(X) is a primitive polynomial according to Definition 4.13.

From example 4.14 we can see that α is a primitive element in $GF(4)$.

4.4.9.2   BCH Codes Construction Using Generator Polynomial

Definition 4.15 *Binary BCH Codes*
For each positive integer $m \geq 3$ and $\leq 2^{m-1}$, there exists a t-error correcting Binary BCH *(n, k t)* code with a block length $(n = 2^m - 1)$ and a number of parity check bits $(n - k \leq mt)$ and a minimum distance $(d \geq 2t - 1)$.

BCH codes can be constructed using a generator polynomial as we will see in this section, but first, we need to introduce the definition of minimal polynomials as follows:

Definition 4.16 *Minimal Polynomial*
A minimal polynomial of an element $a \in GF(2^m)$ with respect to $GF(2)$ is a non-zero monic polynomial $f(x) \in Z_2(x)$, which has the least degree in $Z_2[x]$ S such that $f(x) = 0$. The minimal polynomial has a degree $\leq m$.

Example 4.15 Find the minimal polynomial of the elements of $GF(8)$ that is constructed using the primitive polynomial $x^3 + x + 1$

$$GF(8) = 0, 1, x, x+1, x^2, x^2+1, x^2+x, x^2+x+1.$$

Solutions:

First, we find the elements of this field. $GF(8)$ is constructed using the primitive polynomial $x^3 + x + 1$, which has a root (a), this means

$$a^3 = a + 1$$

The nonzero elements of GF(8) are the powers of a (see Definition 4.16)

$$a^0 = 1$$
$$a^1 = a$$
$$a^2 = a^2$$
$$a^3 = a + 1$$
$$a^4 = a^2 + a$$
$$a^5 = a \times a^4 = a(a^2 + a) = a^3 + a^2 = 1 + a + a^2$$
$$a^6 = (a \times a^5) = a + a^2 + a^3 = 1 + a + a + a^2 = 1 + a^2$$
$$a^7 = (a \times a^6) = a + a^3 = 1 + a + a = 1$$

Now we can identify the corresponding minimal polynomials of $GF(8)$ as follows.

The elements 0 and 1 will have minimal polynomials x and x + 1.

The minimal polynomial for a is the primitive polynomial $x^3 + x + 1$, this polynomial also has two other roots a^2, a^4 (this can be verified by substituting x with a^2 and a^2), take, for example, the case of a^2

$$x^3 + x + 1 = (a^2)^3 + a^2 + 1 = a^6 + a^2 + 1 = a^2 + 1 + a^2 + 1 = 0$$

The remaining three elements of $GF(8)$ a^3, a^5, a^6 satisfy the cubic polynomial $x^3 + x^2 + 1$, so it must be their minimal polynomial. The table below summarises the minimal polynomials associated with each element of $GF(8)$.

Elements	Minimal polynomial
0	x
1	$x+1$
a, a^2, a^4	$x^3 + x + 1$
a^3, a^5, a^6	$x^3 + x^2 + 1$

Now we are ready to define the generator polynomial of a binary BCH code as follows:

Definition 4.17 *Generator Polynomial of Binary BCH Codes*
  Let a be a primitive element in $GF(2^m)$.
  For $1 \leq i \leq t$, let $\theta_{(2i-1)}(x)$ be the minimum polynomial of the field element a^{2i-1}.
  The generator polynomial $g(x)$ of a t-error correcting BCH binary codes of length $2^m - 1$ is given by

$$g(x) = LCM\{\theta_1(x), \theta_3(x)\ldots\theta_{2t-1}(x)\}$$

  LCM is the least common multiple.

Since the degree of each minimal polynomial is m or less, the degree of $g(X)$ is at most mt, this means the number of parity check digits, $n - k$, of the code is at most equal to mt.

Example 4.16 Given the table below of the elements of $GF(2^4)$ (generated by the primitive polynomial $p(x) = x^4 + x + 1$ with their corresponding minimal polynomials. Find the generator polynomial of the Binary BCH code capable of correcting two errors.

Elements	Minimal polynomial
0	x
1	$x + 1$
a, a^2, a^4, a^8	$x^4 + x + 1$
a^3, a^6, a^9, a^{12}	$x^4 + x^3 + x^2 + 1$
a^5, a^{10}	$x^2 + x + 1$
$a^7, a^{11}, a^{13}, a^{14}$	$x^4 + x^3 + 1$

Solution:
  In this case, $m = 4$ and $t = 2$, therefore, $n = 2^4 - 1 = 15$ according to Definition 4.15.
  Following the Definition 4.17, this code can be generated by

$$g(x) = LCM\{\theta_1(x), \theta_3(x)\}$$
$$g(x) = \theta_1(x) \cdot \theta_3(x)$$
$$g(x) = (x^4 + x + 1) \cdot (x^4 + x^3 + x^2 + 1)$$
$$g(x) = x^8 + x^7 + x^6 + x^4 + 1$$

The parity check polynomial $h(x) = \frac{x^{15}+1}{x^8+x^7+x^6+x^4+1}$ according to Definition 4.10 (remember BCH codes are cyclic codes), this gives

$$h(x) = x^7 + x^6 + x^4 + 1$$

The number of data bits k is equal to the degree of the parity check polynomial, therefore $k = 7$.

This means that this is a cyclic code $(n, k, t) = (15, 7, 2)$.

4.4.9.3   BCH Code Construction by Extending Hamming Codes

This is another approach for constructing BCH codes based on extending Hamming codes. First, we will introduce a new definition of Hamming codes.

> **Definition 4.18** *Hamming Codes as Cyclic Codes*
>   Hamming Codes of length n (H_n) are a special class of cyclic codes whose generator polynomials are primitive of degree m such that $n = 2^m - 1$. The dimension of this code is $k = 2^m - m - 1$, where $m \geq 2$.

Recall Example 4.8, where the primitive polynomial ($g(x) = x^2 + x + 1$) of degree $m = 2$ is used to construct a single error correcting code $C = \{000, 111\}$ with $n = 2^m - 1$. Let α be the root of an irreducible polynomial of degree m, this means the powers of α can generate the nonzero elements of the Galois Field $GF(2^m)$ (refer to Definition 4.12 and Example 4.13).

For example, $GF(2^4)^* = \{1, 1+x, x\} = \{\alpha^0, \alpha^1, \alpha^2\}$ (see Example 4.13).

Now if we recall from Sect. 4.4.6 that the parity check matrix of a Hamming code is an $(m \times n)$ matrix whose columns are all non-zero binary m-tuples (i.e. the nonzero elements of the Galois Field $GF(2^m)$). This means we can use the coefficients of the polynomial corresponding to the powers of α as the column vectors of the parity-check matrix, in other words,

$$H = \begin{bmatrix} \alpha^0 & \alpha^1 & \cdots & \alpha^{n-1} \end{bmatrix}$$

Example 4.17 Use the primitive polynomial $x^2 + x + 1$ to construct a parity-check matrix for the binary Hamming code H_3 and deduce its parity check polynomial.

Solution:
The solution to Example 4.13 expresses the powers of a as $\{1, 1+x, x\}$, we use the coefficients of these polynomials as the column vectors of a parity-check matrix, this results in

$$H = \begin{bmatrix} 1 & 1 & 0 \\ 0 & 1 & 1 \end{bmatrix}$$

We can use Definition 4.10 to deduce the parity check polynomial of this code as follows:

$$h(x) = 1 + x$$

Definition 4.10 also states that $h(x).g(x) = 1 + x^n$ we can use this to check our solution. In this case, $h(x).g(x) = (1+x)(1+x+x^2) = 1+x+x^2+x+x^2+x^3$.

As the operation is done over the Galois Field $GF(2^m)$, which is an extension of the binary field Z_2.

Then, $(x+x = 0)$ and $(x^2 + x^2 = 0)$ as this is a binary addition.

This means the above term can be rewritten as:

$$h(x).g(x) = 1 + x^3,$$

which satisfies Definition 4.10.

To be able to correct more errors, we can add more restrictions on the codewords to increase the minimum Hamming distance of the code, hence the number of error corrected.

The BCH codes do this by extending the parity-check matrix:

$$H = \begin{bmatrix} \alpha^0 & \alpha^1 & \cdots & \alpha^{n-1} \end{bmatrix}$$

used to construct Hamming codes by adding extra rows. However, adding extra rows does not necessarily add more restrictions. Assume, for example, we add a second row whose elements are the squares of the corresponding elements of the first row as shown below

$$H = \begin{bmatrix} \alpha^0 & \alpha^1 & \cdots & \alpha^i & \cdots & \alpha^{(n-1)} \\ \alpha^0 & \alpha^2 & \cdots & \alpha^{2i} & \cdots & \alpha^{2(n-1)} \end{bmatrix}$$

This means in this new code $c(x)$ is a codeword if and only if $c(a) = 0 = c(\alpha^2) = 0$

$$(c(a))^2 = \left(a_{n-1}a^{n-1} + \cdots + a_1 a + a_0\right)^2 = \left(a_{n-1}(a^2)^{n-1} + \cdots + a^2 + a_0\right) = c(a^2)$$

Since, on squaring $(c(a))$, the cross terms have coefficient $2 = 0$ and therefore vanish, while the coefficients $a_i = 0, 1$ in $c(x)$ satisfy $a_i^2 = a_i$. Hence, a^2 is a root of $c(x)$ if and only if a is a root, which means adding this extra requirement does not impose any new restrictions on the codewords, hence it is redundant.

Now, if we add a second row whose elements are the cubes of the corresponding elements of the first row as shown below

$$H = \begin{bmatrix} \alpha^0 & \alpha^1 & \cdots & \alpha^i & \cdots & \alpha^{(n-1)} \\ \alpha^0 & \alpha^3 & \cdots & \alpha^{3i} & \cdots & \alpha^{3(n-1)} \end{bmatrix}$$

This is equivalent to extending the columns of H to length $2m$ by adjoining the coefficients of the polynomials representing the cube powers α^{3i}, this means, in this new code $c(x)$ is a codeword if and only if $c(a) = c(\alpha^3) = 0$.

This does really impose further constraints on the codewords. This new code corrects two errors, as follows. Suppose that a codeword c is transmitted, and $b = c + e_k + e_l$ is received, so that there are errors in positions k and l, with $k \neq l$. The receiver computes the syndrome

$$S = bH_T = cH_T + e_kH_T + e_lH_T = e_kH_T + e_lH_T$$

This gives two equations with two unknowns (k, l), the receiver will be able to locate and correct the two errors by solving the equations.

We are now going to generalise this construction of BCH codes as follows:

Definition 19 *Binary BCH Codes*

A Binary BCH code C of a minimum designed distance γ consists of the codewords $c(x)$ of length n such that

$$c(a^b) = c(a^{b+1}) = \cdots = c(a^{b+\gamma-2}) = 0$$

This means the consecutive powers of a are roots of $c(x)$.

a is the generator of the multiplicative group, $F_{2^n}^*$ with a degree m such as $n = 2^m - 1$.

The code C has the following parity-check matrix

$$H = \begin{bmatrix} \alpha^0 & \alpha^b & \alpha^{2b} & \cdots & \alpha^{(n-1)b} \\ \alpha^0 & \alpha^{b+1} & \alpha^{2(b+1)} & \cdots & \alpha^{(n-1)(b+1)} \\ \vdots & \vdots & \vdots & \vdots & \vdots \\ \alpha^0 & \alpha^{b+\gamma-2} & \alpha^{2(b+\gamma-2)} & \cdots & \alpha^{(n-1)(b+\gamma-2)} \end{bmatrix},$$

where each element represents a column-vector of length m.

if $b = 1$, then the BCH code is called a narrow sense BCH code in this case,

$$H = \begin{bmatrix} \alpha^0 & \alpha^1 & \alpha^2 & \cdots & \alpha^{(n-1)} \\ \alpha^0 & \alpha^2 & \alpha^4 & \cdots & \alpha^{2(n-1)} \\ \vdots & \vdots & \vdots & \vdots & \vdots \\ \alpha^0 & \alpha^{\gamma-1} & \alpha^{2\gamma} & \cdots & \alpha^{(n-1)(\gamma-1)} \end{bmatrix}$$

Now the Hamming distance for this code $d \geq \gamma$, in many cases $d > \gamma$.

Example 4.18 The binary Hamming codes are BCH codes with $b = 1$ and designed distance $\gamma = 2$, but in fact, they have minimum distance $d = 3$.

Example 4.19

(a) Use the primitive polynomial $1 + x + x^4$ to construct a parity-check matrix for the Binary Hamming code H_{15}.
(b) Then extend this matrix to construct the parity-check matrix for a narrow sense Binary BCH with a designed distance $\gamma = 4$.

Solution:

In this example, the degree of the primitive polynomial is 3, therefore $m = 3$ and $n = 2^3 - 1 = 15$.

The parity-check matrix for the corresponding Hamming code will have the following form:

$$H = \begin{bmatrix} \alpha^0 & \alpha^1 & \cdots & \alpha^{15} \end{bmatrix}$$

a is the root of the primitive polynomial $1 + x + x^4$, where the elements of this matrix correspond to the coefficients of the polynomial corresponding to the powers of α.

Therefore, we can write

$$1 + a + a^4 = 0$$
$$a^4 = a + 1$$

We can now find the polynomials corresponding to the powers of a by substituting a^4 using the above equation as follows:

$$a^0 = 1$$
$$a^1 = a$$
$$a^3 = a^3$$
$$a^4 = a + 1$$
$$a^5 = a \times a^4 = a(a + 1) = a^2 + a$$

Similarly, we find:

$$
\begin{aligned}
a^6 &= a^3 + a^2 \\
a^7 &= a^3 + a + 1 \\
a^8 &= a^2 + 1 \\
a^9 &= a^3 + a \\
a^{10} &= a^2 + a + 1 \\
a^{11} &= a^3 + a^2 + a \\
a^{12} &= a^3 + a^2 + a + 1 \\
a^{13} &= a^3 + a^2 + 1 \\
a^{14} &= a^3 + 1 \\
a^{15} &= a^4 + a = 1 + a + a = 1
\end{aligned}
$$

We can now use the coefficients of the above polynomials corresponding to the powers $a^0, a, a^2, \ldots, a^{14}$ to construct the following parity-check matrix. The resulting matrix is given below.

$$
H = \begin{bmatrix}
1 & 0 & 0 & 0 & 1 & 0 & 0 & 1 & 1 & 0 & 1 & 0 & 1 & 1 & 1 \\
0 & 1 & 0 & 0 & 1 & 1 & 0 & 1 & 0 & 1 & 1 & 1 & 1 & 0 & 0 \\
0 & 0 & 1 & 0 & 0 & 1 & 1 & 0 & 1 & 0 & 1 & 1 & 1 & 1 & 0 \\
0 & 0 & 0 & 1 & 0 & 0 & 1 & 1 & 0 & 1 & 0 & 1 & 1 & 1 & 1
\end{bmatrix}
$$

In order to extend this matrix to build a Binary BCH with a designed distance of 4, we will need to construct the following matrix according to Definition 4.19:

$$
H = \begin{bmatrix}
\alpha^0 & \alpha^1 & \cdots & \alpha^{14} \\
\alpha^0 & \alpha^2 & \cdots & \alpha^{28} \\
\alpha^0 & \alpha^3 & \cdots & \alpha^{42}
\end{bmatrix}
$$

From the above calculation of the powers of a we note that the powers of a repeat with a period of 15, for example, $a^{16} = a^{15}.a^1 = a^1$, we can use this to find all powers up to $a^{42} = a^{15}.a^{15}.a^{12} = a^{12} = a^3 + a^2 + a + a + 1$.

From the previous discussion, we noted that the second row of the matrix above (i.e. the square is redundant), so no need to be included in the matrix.

This allows us to construct the required parity-check matrix by replacing the powers of a with the coefficients of their corresponding polynomials, as follows:

$$H = \begin{bmatrix} 1 & 0 & 0 & 0 & 1 & 0 & 0 & 1 & 1 & 0 & 1 & 0 & 1 & 1 & 1 \\ 0 & 1 & 0 & 0 & 1 & 1 & 0 & 1 & 0 & 1 & 1 & 1 & 1 & 0 & 0 \\ 0 & 0 & 1 & 0 & 0 & 1 & 1 & 0 & 1 & 0 & 1 & 1 & 1 & 1 & 0 \\ 0 & 0 & 0 & 1 & 0 & 0 & 1 & 1 & 0 & 1 & 0 & 1 & 1 & 1 & 1 \\ 1 & 0 & 0 & 0 & 1 & 1 & 0 & 0 & 0 & 1 & 1 & 0 & 0 & 1 & 0 \\ 0 & 0 & 0 & 1 & 1 & 0 & 0 & 0 & 1 & 1 & 0 & 0 & 0 & 0 & 1 \\ 0 & 0 & 1 & 0 & 1 & 0 & 0 & 1 & 0 & 1 & 0 & 0 & 1 & 1 & 0 \\ 0 & 1 & 1 & 1 & 1 & 0 & 1 & 1 & 1 & 1 & 0 & 1 & 1 & 1 & 1 \end{bmatrix}$$

4.4.9.4 Decoding BCH

In this section, we are going to explain the procedure of BCH decoding using an example case. Consider the binary BCH code $(n, k, t) = (15, 7, 2)$, defined over Galois field $GF(2^4)$ based on the primitive polynomial $p(x) = 1 + x + x^4$. The generator polynomial $g(x)$ of this code is given as (see Example 4.16)

$$g(x) = \left(x^4 + x + 1\right) \cdot \left(x^4 + x^3 + x^2 + 1\right)$$
$$g(x) = x^8 + x^7 + x^6 + x^4 + 1$$

Assume a codeword $y(x)$ is received which has two errors, which means it has a Hamming distance of 2 from the closest valid codeword. Assume the positions of these two errors are i_1, i_2.

Now let

$$y(x) = \sum_i y_i X^i$$

And the closest valid codeword is

$$c(x) = \sum_i c_i X^i$$

Also, the error vector is:

$$e(x) = X^{i1} + X^{i2}$$

By definition:

$$y(x) = c(x) + e(x) = d(x)g(x) + e(x) \text{ where } d(x) \text{ is the data word.}$$

Therefore:

$$y(x) = e(x) \text{ Whenever } x = \text{any root of } g(x)$$

Remember: $g(x)$ is a divisor of all valid codewords $c(x)$ and the roots of $g(x)$ are also roots of $c(x)$.

We already know that $a, a^2, a^4,\ a^8,\ a^3, a^6, a^9$ and a^{12} are roots of $g(x)$ (see Example 4.16).

Suppose the errors occur at locations $i_1, i_2 \ldots i_v$, we define the syndrome equations as follows:

$$S_j = y(a^j) = \sum_{i=1}^{v} (a^j)^i = \sum_{i=1}^{v} (a^i)^j$$

Now let

$$X_i = a^i$$

The syndrome equation can be rewritten as follows:

$$S_j = \sum_{i=1}^{v} X_i^j$$

In this case, we only have two errors on locations i_1, i_2.
Therefore, we will only have two variables

$$X_{i1} = a^{i1}$$
$$X_{i2} = a^{i2}$$

Typically we will need 2t equations to locate t errors, in this case, t = 2, therefore, we need four equations. We will consider the equations corresponding to the first four powers of a are listed below

$$S_1 = y(a) = X_{i1} + X_{i2}$$
$$S_2 = y(a^2) = X_{i1}^2 + X_{i2}^2$$
$$S_3 = y(a^3) = X_{i1}^3 + X_{i2}^3$$
$$S_4 = y(a^4) = X_{i1}^4 + X_{i2}^4$$

The equations are said to be power sum symmetric functions. Our task is to find X_{i1} and X_{i2}, two non-zero and different elements in GF(16) satisfying the above equations.

One approach is to find the product $X_{i1}X_{i2}$ from these equations. In which case, we will have knowledge of $X_{i1} + X_{i2}$ (which is S_1) and $X_{i1}X_{i2}$, so that we can construct the polynomial

$$(X_{i1} - z)(X_{i2} - z) = z^2 - (X_{i1} + X_{i2})z + X_{i1}X_{i2}$$

Factoring this polynomial then reveals X_{i1} and X_{i2}
We note that:

$$S_1^3 = (X_{i1} + X_{i2})^3$$
$$= (X_{i1}^2 + X_{i2}^2)(X_{i1} + X_{i2})$$
$$= X_{i1}^3 + Y^3 + X_{i1}X_{i2}(X_{i1} + X_{i2})$$
$$= S_3 + X_{i1}X_{i2}S_1$$

Since $S_1 \neq 0$, we find that $X_{i1}X_{i2} = S_1^2 - S_2/S_1$, and we can thus decode the code.

Example 4.19 Consider the [15, 7, 2] BCH code constructed based on $GF(2^4)$ generated by the primitive polynomial $p(x) = x^4 + x + 1$ from Example 4.16 with a generator polynomial:

$$g(x) = (x^4 + x + 1) \cdot (x^4 + x^3 + x^2 + 1)$$

Decode the received codeword below:

$$y(x) = x^4 + x^6 + x^7 + x^8 + x^{13},$$

which corresponds to:

$$y = (0,0,0,0,1,0,1,1,1,0,0,0,0,1,0).$$

Solution:
Let us first recall the powers of a

$$a^0 = 1$$
$$a^1 = a$$
$$a^3 = a^3$$
$$a^4 = a + 1$$
$$a^5 = +a$$
$$a^6 = a^3 + a^2$$
$$a^7 = a^3 + a + 1$$

$$a^8 = a^2 + 1$$
$$a^9 = a^3 + a$$
$$a^{10} = a^2 + a + 1$$
$$a^{11} = a^3 + a^2 + a$$
$$a^{12} = a^3 + a^2 + a + 1$$
$$a^{13} = a^3 + a^2 + 1$$
$$a^{14} = a^3 + 1$$

We already know that $a, a^2, a^4, a^8, a^3, a^6, a^9$ and a^{12} are roots of $g(x)$, let us first write the syndrome equations for the first four powers of a by substituting the first four powers of a in equation $y(x)$ (remember $a^4 = a + 1$), we can compute the following syndrome equations:

$$S_1 = y(a) = a^3 + a^2 = a^6$$
$$S_2 = y(a^2) = a^3 + a^2 + a + 1 = a^{12}$$
$$S_3 = y(a^3) = a^3 + a + 1 a^7 = a^7$$
$$S_4 = y(a^4) = a^3 + a = a^9$$

The polynomial to be factored is $z^2 + (a^3 + a^2)z + a^{12} - a = z^2 + a^6 z + a^{13}$. There are various ways to factor this polynomial, one of which is to try all possible values of a^i (this search is called Chien Search) [8, 9], which gives the following solutions a^1, a^{13}.

This means the received codeword has errors at positions 0 and 13, so the closest codeword is given as follows:

$$Y = (1, 0, 0, 0, 1, 0, 1, 1, 1, 0, 0, 0, 0, 0, 0).$$

4.4.10   Block Error Probability

The block error probability refers to the probability of a decoding error (i.e. the decoder picks the wrong codeword when applying the Hamming distance rule). For a t-error correcting code of length n, this probability is given as follows:

$$P_{ECC} = 1 - \sum_{i=0}^{t} \binom{n}{i} P_e^i (1 - P_e)^{n-i}, \tag{4.4}$$

where P_e is the expected bit error rate (as explained in Sect. 4.3).

This formula can be used to determine the length of the error correction code (n) to be used given the bit error rate of the PUF response and the desirable

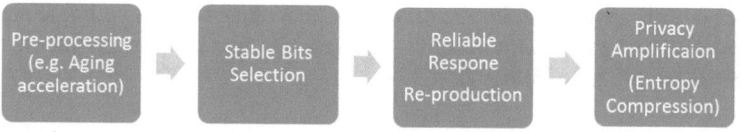

Fig. 4.4 A generic flow of PUF reliability enhancement techniques

decoding error (P_{ECC}), the latter is typically in the order of 10^{-6} (see Chap. 6 for a detailed example on how to use the above formula to select an appropriate error correction code given a specific bit error rate).

4.5   Reliable PUF Design Flow

The aim of this flow is to reduce the bit error rate of the PUF responses below a specific level set by the intended application. For example, cryptographic key generation schemes require precisely reproducible responses, whereas authentication protocols may be more tolerant to noise-induced errors. Figure 4.4 shows a generic design flow of reliable PUF circuits.

The first stage is an optional pre-processing step in which the PUF chip is purposely aged in order to limit progressive change in their output in the field [10, 11]. In the second stage, the least reliable bits of the PUF response are identified and discarded, this stage is also not mandatory but it can help reduce the hardware cost of the third stage. The latter reduces the error rate to an acceptable level using generic fault tolerance schemes, such as temporal redundancy (e.g. using temporal majority voters), information redundancy (error correction codes) or hardware redundancy. The response obtained from the third stage does not typically have acceptable entropy from a security perspective because of error correction and the inherent biases of the PUF architecture. Therefore, a fourth stage may sometimes be required to increase the entropy of the PUF responses [12]. Examples of privacy amplifiers include hash functions and encryption cores. We will now give an in-depth insight into the existing approaches applied at the first three stages of the flow in Fig. 4.2.

4.6   Reliability Enhancement Using Pre-processing Techniques

The essence of these methods is to exploit the IC aging phenomenon (see Chap. 3 for more details) to reinforce desired PUF responses by permanently altering the electrical characteristics of the PUF circuits; such reinforcement can help improve the stability of the PUF responses and reduce its expected bit error rate. There are a number of examples in the literature of how such approach can be implemented in practice. We will study two of such approaches, the first is based on the use of BTI and the second is based on HCI.

4.6.1   BTI Aging Acceleration

We are going to use an SRAM PUF as an example to explain how this technique can be applied for a SRAM cell (see Fig. 4.5) pull-up transistors, P1 and P2 suffer from NBTI. The pull-down transistors (N1, N2) are affected by PBTI. P1 and P2 transistors are part a cross-coupled inverter, this means only one would be under NBTI stress at any moment in time.

Such asymmetric stress conditions result in unbalanced threshold voltage degradation of these two transistors, for example, if P1 is placed under greater stress than P1, its threshold voltage degradation will be bigger compared to P2, which will reduce its driving capability, consequently, node Q is less likely to be pulled up to a

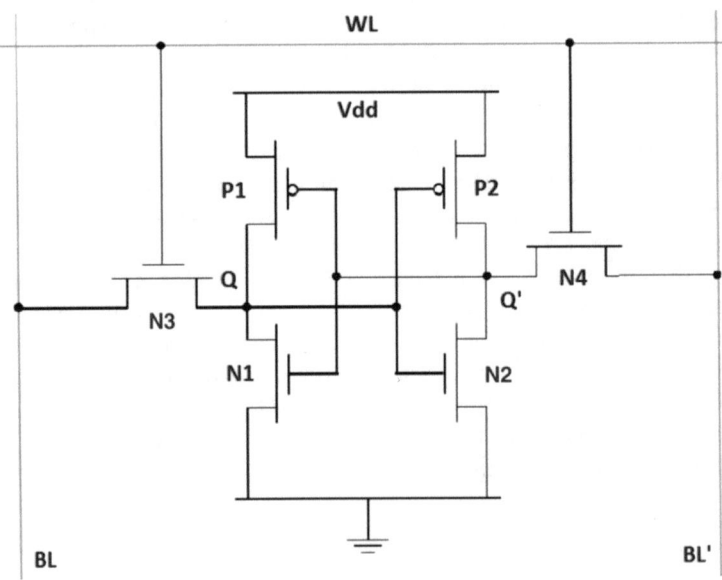

Fig. 4.5 A generic 6-transistor SRAM cell

logic 1 once the cell is powered up. This effect can be exploited to reinforce a preferred value in the SRAM cell since the polarity of the resolved bit is a strong function of the relative threshold voltages of the devices. This can be explained as follows, each SRAM cell has a preferred initial state that it assumes when powered up wherein the PMOS device with a lower threshold voltage is turned on. If we are to store the opposite of this preferred value, the PMOS device with higher Vth would be turned on, this means its driving capability gets weaker and its Vth increases due to NBTI stress, which means the threshold voltage difference between the two PMOS devices increases. This effectively means the SRAM cell will be more capable to assume the same response every time it is powered up, which makes the SRAM PUF more reliable.

An example of such effects is [13], wherein the authors have employed the static noise margin to quantify the impact of BTI on SRAM.

Aging acceleration is not only applicable to SRAM PUF, in fact, the same technique can be used to enhance the reliability of other PUF designs such as delay-based architectures (e.g. Ring Oscillator PUF). This is because the rate of BTI aging degradation is typically larger for fresh circuits, as the circuits age its degradation slows downs [14, 15], which means the response of a PUF circuit is unlikely to change after it has been aged. Therefore, one can deduce the following generic procedure for this approach:

1. Establish a preferred response of a PUF, by reading it multiple times.
2. Identify the best state of the PUF to reinforce the preferred response (only needed in memory-type PUF, such as SRAM or latch-based PUF).
3. Induce accelerated BTI aging by increasing temperature(e.g. place the device in a burn-in oven) or using high supply voltages.
4. Characterise the PUF circuit to identify its new post-aging challenge/response behaviour.

The use of NBTI aging acceleration has been estimated to achieve a sizable reduction in the expected bit error rate of PUFs, the authors of [16] reported a 40% improvement in the reliability of an SRAM PUF using this approach. In a more recent study [11], Satpathy et al. reported a 22% decrease in the BER of an SRAM PUF as a result of direct acceleration of BTI aging to reinforce the pre-existing bias in the PUF cells and to inject a bias into the clock delay-path by aging the buffers and the pre-charge transistors.

Despite its clear benefits, the BTI aging acceleration approach has a number of drawbacks, first of all, the high temperature-based acceleration cannot be applied selectively, therefore, other circuits on the chip would also age. More importantly, BTI degradation is not permanent, as aging effects are partly reversed if the device is placed in a recovery state [14]. There are a number of methods to overcome such disadvantages. For example, the use of voltage-based aging acceleration can help avoid the aging the whole chip by applying high voltage pulses selectively to the PUF circuit. In addition, increasing the stress time can help achieve a permeant shift in the threshold voltage [17].

4.6.2   HCI Aging Acceleration

HCI is another aging mechanism, wherein traps are formed in the gate oxide of a CMOS device. These traps are generated when channel carriers are accelerated to gain a sufficient energy level to allow them to penetrate into the oxide. HCI mainly affects NMOS devices as their charge carriers (i.e. electrons) has higher mobility than those of the PMOS transistors (holes). Unlike BTI aging, HCI-induced degradation is more permanent [18], therefore, it may produce more stable PUF responses. An example on the use of this approach is reported in [10], wherein authors incorporated additional control circuitry to each PUF architecture, which is capable of inducing HCI stress conditions in the PUF device without affecting the surrounding circuitry. This is achieved by applying a series of high voltage short pulses onto selected transistors, which lead to an increase in the flow of high energy current carriers over the CMOS channels, some of these will cause the formation of traps in the gate oxide of the associated device and causes a permeant HCI-induced threshold voltage shift. The authors reported a 20% decrease in the expected bit error rate of the PUF response; this comes at the expense of a 10 folds increase in the area of the PUF device due to the additional control circuitry. The use of HCI to improve the reliability of PUF responses is an interesting concept but incurs significant area overheads, therefore, more work is still needed to establish its efficacy in practice.

4.7   Reliability Enhancement Using Stable Bits Selection

The essence of this technique is to discard the least reliable bits in the PUF responses, which reduces the need for expensive error correction circuitry. There are many ways to implement such a technique, this primarily depends on the specific features of the PUF architecture and the type of errors to be reduced [12]. In general, this approach consists of two stages; first, the bit error rate associated with each response bit is estimated (i.e. $P_e(r_i)$), then a subset of the total response bits are chosen, such that their respective error rates are below a certain threshold value. A conceptual diagram of this method is shown in Fig. 4.6, where $\{r_1 \ldots r_n)$ represent the original output bits of the PUF device and $\{r_q \ldots r_m)$ represent the selected stable bits.

We are going to discuss two examples of bit selection schemes, namely; index bit masking (IBS) and stable-PUF-marking (SPM).

4.7.1   Index Based Masking

This approach was first proposed in [12] to reduce impact of environment noise (e.g. Temperature and power supply variations) on ring oscillator PUFs. To explain this method, let us consider the generic structure of an RO PUF depicted in Fig. 4.7.

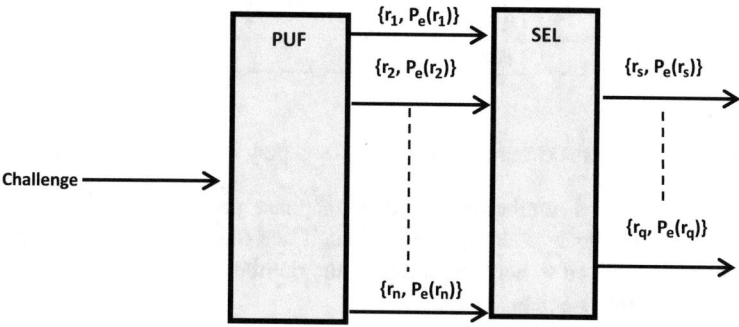

Fig. 4.6 Generic diagram for stable bits selection schemes

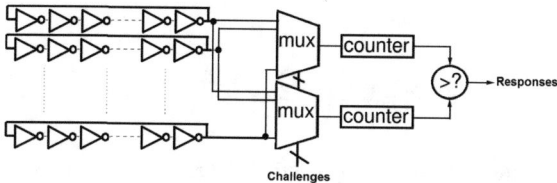

Fig. 4.7 Generic structure of ring oscillator PUFs

It consists of two multiplexers, two counters, one comparator and K ring oscillators [19]. Each ring oscillates at a unique frequency depending on the characteristics of each of its inverters, which vary from one cell to another due to manufacturing variations, the two multiplexers select two ROs to compare. The two counter blocks count the number of oscillations of each of the two ROs in a fixed time interval. At the end of the interval, the outputs of the two counters are compared and depending on which of the two counters has the highest value, the output of the PUF is set to 0 or 1.

This architecture is very sensitive to temperature and power supply variations, which affects the delay of the inverter chains, hence their oscillation frequencies. This can cause bit flips in the PUF response if the change in the original oscillation frequency is sufficiently significant. To avoid such a scenario, a designer can preselect a subset of ring oscillators to be compared, such that the difference of their original frequencies is larger than any possible frequency shift due to noise or environment variations. This approach implies that the number of available challenge/response pairs will be reduced, as the number of ring oscillator's pairs available for comparison is decreased. To illustrate this, we will give a numerical example.

Example 4.20 Let us consider a ring oscillator PUF that has four inverter chains whose respective frequencies are given in the table below.

Chain	R1	R2	R3	R4
Frequency (MHz)	100	106	97	110

Assume that the maximum variation in frequency due to noise is ±2 MHz.

1. Calculate the total number of challenge/response pairs for this PUF.
2. IBS approach was applied to this PUF such that the number of bit flips caused by noise is reduced to zero. Recompute the number of challenge/response pairs for this reliability-enhanced PUF.

Solution:

1. The number of challenge/response pairs for RO PUF with $k = 4$ inverter chains can be calculated as follows:

$$CRP = \frac{k.(k-1)}{2} = \frac{4.(4-1)}{2} = 6$$

2. When IBS is applied, only ring oscillator pairs whose respective frequencies differ by more than 5 MHz can be compared, this is to ensure no error will occur even in the worst-case scenario, i.e. when the frequencies of two oscillators shift in opposite directions.

This gives us a set of four allowed pairs: {(R1,R2), (R1,R4), (R2,R3) and (R3,R4)}
The forbidden pairs are {(R1,R3),(R2,R4)}, which have a frequency difference of 3 and 4 MHz, respectively.
Therefore, in this case, the number of challenge/response pairs is reduced to 4.
This preselection can be performed on-chip using additional control logic. The authors of [12] presented a soft decision-based implementation of the selection block, where indices are used to select the bits that are less likely to be noisy. Such an approach does incur extra area overhead and it is important that such increase in silicon costs does not exceed any potential area savings obtained by using a lighter error correction code.

4.7.2   Stable-PUF-Marking

This is another stable bits selection scheme that was proposed for SRAM PUFs in [20]. This approach consists of two stages: the purpose of the first stage is to identify the most reliable cells in a SRAM PUF, in the second stage these cells are marked as stable, and the responses of the remaining unreliable cells are discarded. This process takes place at the post-fabrication stage, where each PUF chip is characterised and marked.

The challenge for this approach is how to identify stable cells, the most intuitive approach is to perform multiple readings at different environmental conditions and choose the bits that are most stable, however; this requires extensive chip testing, which may not be feasible given the large volume of devices and strict time constraints in industrial chips production facilities. The authors in [20] proposed an alternative approach to reduce the number of tests required to identify stable chips, their solution exploits the cross-coupled transistors architecture present in the SRAM cell. When the latter is powered up, the cell will assume a value of "0" or "1" determined by the inherent difference in the driving capabilities of the cross-coupled transistors (see Fig. 4.5 for a SRAM cell diagram). Crucially, the larger the mismatch between these transistors, the faster the cell resolves to a stable value. Therefore, if one can measure the time each cell takes to resolve to a stable state, one can use this as an accurate indicator on the degree of mismatch. This fact can be exploited to determine the cells whose mismatch is larger than a certain threshold; indeed, the authors of [20] demonstrated that stable cells can be identified with only two readings per SRAM cell which significantly reduces the overhead incurred by time-consuming exhaustive testing.

4.8   Reliable PUF Response Generation Methods

Although pre-processing techniques and reliable bit selection schemes can help improve the stability of PUFs responses, a post-processing stage is typically needed to ensure the device operates correctly when deployed in the field. There are a number of approaches for the generation of reliable bits responses, which are based on conventional fault tolerance schemes: information redundancy, hardware redundancy and temporal redundancy. In the first approach, a designer would need to embed additional information into the device, normally referred to as "Helper Data" in order to help correct potential noise-induced bit flips. In the second approach, the PUF design area is extended to improve resiliency to errors. In the third approach, some form of majority voting scheme is employed. Examples of these techniques will be explained in the following subsections.

4.8.1   Secure Sketch Using Code-Offset Construction (Information Redundancy)

A secure sketch scheme is a process whereby a noisy version of a PUF response (r′) can be recovered to its original value (r) using a helper data vector (w), it consists of two randomised procedures "Sketch" for helper data generation and "Recover" for response reconstruction, it is based on the use of error correction codes [21].

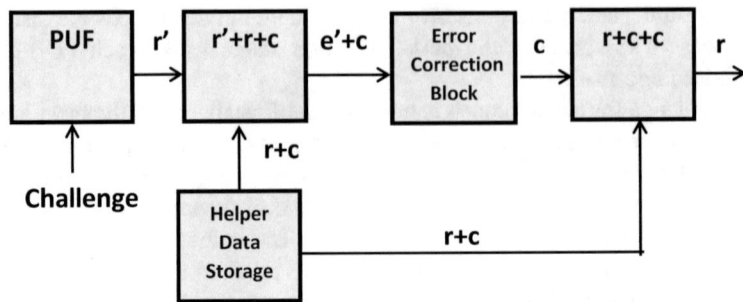

Fig. 4.8 PUF response reproduction using code-offset construction

Code-offset construction consists of two phases; an enrolment phase followed by a response construction stage.

The enrolment phase takes place at the post-fabrication stage before field deployment of PUF devices and consists of the following steps:

1. Identify a suitable error correction codes $C(n, k, t)$, such that n is equivalent to the length of the PUF response r.
2. For each PUF device, generate a set of response/challenge pairs.
3. For each response, add a random codeword $c \in C$ to generate a helper data vector unique to each challenge/response pair $h = r + c$.
4. Store the helper data vectors in the helper Data storage.

The second stage reconstructs the PUF responses using helper data vectors as follows:

1. The PUF generates a response r′, which normally has an additive error $r' = r + e$.
2. The helper data is then Xored with the erroneous response to generate:

$$h + r' = r + c + r = c + r' + r = c + e'$$

3. If the Hamming distance between r and r' $(wtH (e') = dH (r, r'))$ is within the error correction capabilities of the code $C(n, k, t)$, it will be possible to reproduce the original PUF response (r), by first decoding $(e' + c)$ then, XORing the result with the helper data (h).

A block diagram of the response construction phase is shown in Fig. 4.8.

Helper data can either be stored on the same chip as the PUF or they can be sorted off-chip or supplied to the PUF with the challenge. This very much depends on the type of application, for schemes used in device authentication, helper data can be stored on the authentication authority server and transmitted to the PUF as an addition to the challenge bits.

4.8.2   Secure Sketch Using Syndrome Coding Construction

This is another response reproduction method based on the use of error correction codes [21]. It also consists of two phases; an enrolment phase followed by a response construction stage. The enrolment phase takes place before field deployment of PUF devices and consists of the following steps:

1. Identify a suitable error correction code $C(n, k, t)$ with a parity-check matrix H, such that n is equivalent to the length of the PUF response r.
2. For each PUF device, generate a set of response/challenge pairs.
3. Each response can be viewed as a codeword with additive noise $r = c + e$.
4. Generate a syndrome vector: $s = r \cdot H_T$ is the transpose matrix of H, this gives: $s = r \cdot H_T = c \cdot H_T + e \cdot H_T = e \cdot H_T$ (remember $c \cdot H_T = 0$ by definition).
5. Store all syndrome vectors as helper data vectors ($s = e \cdot H_T$) in the helper data storage.

The second stage reconstructs the PUF responses using helper data vectors as follows:

1. The PUF generate response r'.
2. The generated response can also be viewed as a codeword with an additive noise:

$$r' = c + e'$$

This vector can be written as follows:

$$r' = c + e' = r + e + e'$$

3. Generate its syndrome vector: $s' = r' \cdot H_T$, which can be written as follows:

$$s' = r' \cdot H_T = (c \oplus e') \cdot H_T = e' \cdot H_T$$

4. Add the generated syndrome to the corresponding helper data to generate a new syndrome vector $s'' = s' + s = e' \cdot H_T + e \cdot H_T = (e + e')H_T$.
5. If the Hamming distance between r and r' ($wtH (e + e') = dH (r, r')$) is within the error correction capabilities of the code $C(n, k, t)$, i.e. ($wtH (e + e') \leq t$), it will be possible to reproduce the original PUF response (r), by identifying the error vector $(e + e')$ first then XORing the result with the response r'.

A block diagram of the response construction phase is shown in Fig. 4.9.

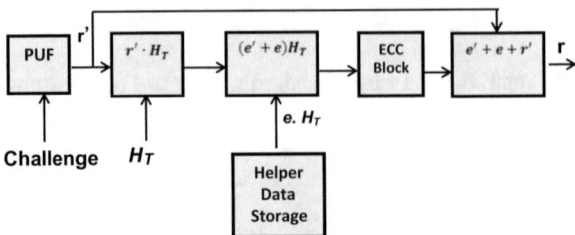

Fig. 4.9 PUF response reproduction using syndrome coding

4.8.3 Temporal Majority Voting

This is another reliable response reproduction method, which is based on the use of time redundancy techniques wherein, an odd number of responses are generated for the same challenge at different time intervals (t_1, t_2 ... t_q), these are initially stored in an on-chip memory. After sufficient number of responses are collected, they are applied to a self-checking majority voter to choose the most reliable response as depicted in Fig. 4.10 [11, 22]. As can be seen from the circuit diagram, this technique requires additional hardware for response storage and control circuity.

This technique helps reduce error caused by transient failure mechanisms such as temporal variations in voltage supply or radiation hits, however, error caused by permanent physical failure mechanisms such as aging cannot be corrected, therefore, error correction may still be necessary. Equation (4.5) computes the error rate ($P_e(out)$) at the output of voters given an error rate ($P_e(in)$) at the input

$$P_e(out) = 1 - Bin_q\left(\frac{q-1}{2}\right) \leq P_e(in), \tag{4.5}$$

where

Bin_q is the cumulative distribution function of the binomial distribution.

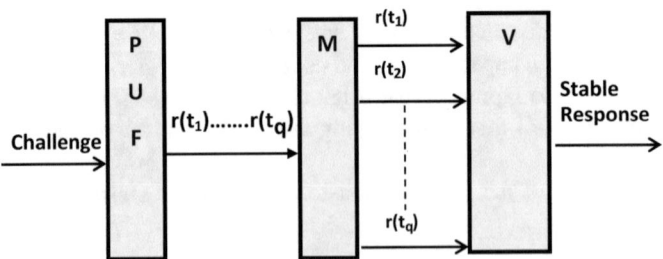

Fig. 4.10 PUF response reproduction using temporal redundancy

Fig. 4.11 A configurable inverter chain

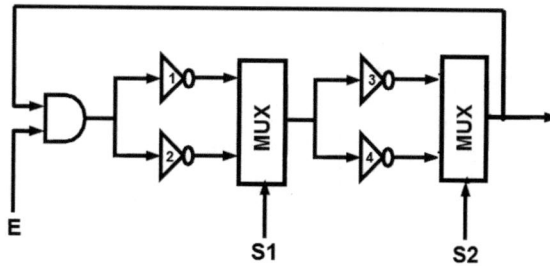

4.8.4   Hardware Redundancy

The essence of this approach is to include extra circuitry to the PUF architecture to improve its resilience against noise. One example of such an approach was proposed in [1], wherein the design of the RO PUF was considered. The authors proposed a scheme wherein each inverter chain is replaced by a configurable structure which allows for multiple paths through the chain, for example, if we have two configuration bits we can obtain four possible chains as depicted in Fig. 4.11.

The frequencies of the configurable chains are then compared and the configurations that produce the largest difference between each respective pair is stored as part of the PUF challenge. This ensures that a maximum possible frequency gap is obtained between each two inverter chains which makes the PUF response more stable.

4.9   Cost Analysis of Reliability Enhancement Approaches

The cost of reliability enhancement techniques is divided into two categories; those incurred by the extra procession steps at the post-fabrication stage and those associated with the extra circuitry needed for a reliable response generation. The latter is the more critical component in resources constrained devices such as RFID tags.

The cost of error correction-based schemes is proportional to the number of errors to be corrected. We are going to consider BCH codes as an example of such a trend as these codes are one of the most widely used schemes for PUF cryptographic key generation [23]. The area overhead of BCH decoding blocks with different error correction capabilities is summarised in Fig. 4.12, based on area estimation reported in [24] using a 65 nm CMOS technology.

The trend shown in Fig. 4.12 indicates that the cost of on-chip reliable response generation increases significantly with the number of correctable errors (t). The data shows that relationship between (t) and area cost is linear, this allows extrapolation of the area costs for $t > 7$.

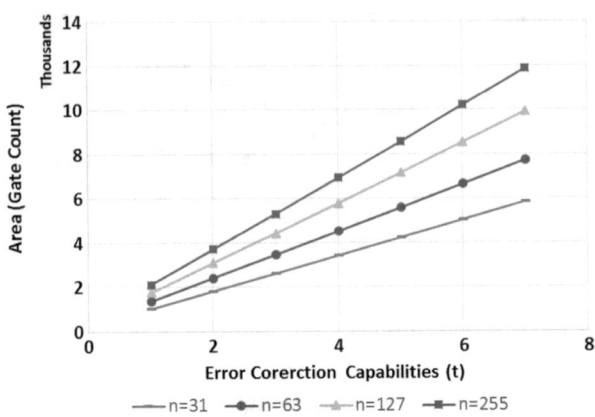

Fig. 4.12 The area costs of the BCH decoders (n—the code length, t—correctable errors)

Table 4.3 Error correction requirements of a PUF-based key generation scheme

Reliability enhancement technique	None	BTI aging acceleration	BTI aging acceleration and soft dark-bit masking	BTI aging acceleration and soft dark-bit masking and temporal majority voting
Achieved reduction in bit error rate	0	1.4%	5%	7.5%
Error correction requirement (t)	22	18	9	2

This may render PUF-based security solutions prohibitively expensive for resources-constrained systems and not even viable for low-end IoT devices [25]. This makes it essential in such cases to reduce the on-chip error correction requirements, which can be achieved using both pre-processing and bit selection schemes at the expense of additional chip processing before deployment. *The question is: how efficient are pre-processing approaches in reducing the requirements of error correction schemes.* This question has been partly answered by a study published by Intel labs [11] wherein the authors developed a stable SRAM PUF key generation scheme. The bit error rate of the PUF without any reliability enhancement has been estimated to be 8.5%. They use BTI aging acceleration, bit selection scheme (Soft dark-bit Masking) and temporal majority voting to reduce this error rate to 1%. Table 4.3 summarises the techniques used and the associated error correction requirements (in term of t). The latter is calculated for n = 256 BCH code needed in each case to reduce the error rate to zero, for example, if the raw PUF output is used, a BCH ECC capable of correcting 22 errors will be needed to ensure reliable responses need for the cryptographic key generation.

The results in Table 4.3 show that the error correction requirement can be decreased by more than twice using pre-processing approaches. This can be further reduced by more than four times if on-chip temporal majority voting is employed. This decrease in (t) implies an equivalent reduction the area associated with the BCH decoders (see sure Fig. 4.12).

Although the use of pre-processing approaches may incur additional costs at the design and fabrication time, the analysis above shows that these techniques are very effective in making the use of PUF technology a viable option for resource-constrained systems.

The area and power costs of error correction schemes can be further decreased using codes in parallel if the response of the PUF is long. This is achieved by dividing the response into non-overlapping separate sections and decoding each part separately. The partial coding approach has been adopted in [26] for low-cost reliable PUF responses generation, it has also been proved cost-effective for other area constrained reliable systems [27].

4.10   Conclusions

Silicon-based PUFs are susceptible to temporal and permanent failure mechanisms, this may lead to a change in their response/challenge behaviours, which can undermine their use in a number of applications such as key generation and device authentication. Reliability-enhancement techniques for PUFs can be classified into two types: pre-procession methods and response reconstruction techniques. Pre-processing techniques take place at the post-fabrication stage; they include a range of approaches such as aging acceleration and reliable bit selection, which aim to increase the robustness of the design before deployment. These approaches can certainly help mitigate the impact of predictable failure mechanisms such as aging, but they may not be able to protect against unpredictable noise such as unexpected fluctuations in environment conditions. Therefore, there is a need for on-chip reliable response construction schemes, which aim to recover the original PUF response from a noisy version, these techniques rely on conventional fault tolerance approaches such as hardware redundancy, time redundancy and information redundancy.

In practice, a combination of the above-mentioned methods need to be implemented to reduce the overall costs.

4.11 Problems

4.1. Which of the following binary codes are linear?

$$C_1 = \{00, 01, 10, 11\}$$
$$C_2 = \{000, 011, 101, 110\}$$
$$C_3 = \{00000, 01101, 10110, 11011\}$$
$$C_5 = \{101, 111, 011\}$$
$$C_6 = \{000, 001, 010, 011\}$$
$$C_7 = \{0000, 1001, 0110, 1110\}$$

4.2. What is the basis of $C_n = \{00...0_n, 11...1_n\}$ and what is the rate of this code? Assuming $n = 7$, how many errors can this code correct?

4.3. You are given the irreducible polynomial $r(x) = x^3 + x + 1$ defined over Z_2

   (a) Construct a Galois field GF (2^3) and find its elements using $r(x)$.
   (b) Is $r(x)$ primitive and why?

4.4. A cyclic code C defined over a ring $R_6 = Z[X]/(X^6 + 1)$ has the generator polynomial $g(x) = 1 + x + x^2$:

   (a) Compute its parity check polynomial $h(x)$.
   (b) Verify whether or not the following are codewords:

$$r1(x) = x^3 + x^4 + x^5$$
$$r2(x) = 1 + x^4$$

4.5. Construct a syndrome computing circuit for a cyclic code with $g(x) = 1 + x + x^3$. Then, using your circuit verify whether or not the following received word is a valid codeword: $r(x) = 1 + x^2 + x^5 + x^6$.

4.6. Construct a Hamming code with length $n = 3$.

4.7. Use the primitive polynomial $1 + x + x^4$ to construct a parity-check matrix for the binary Hamming code H_{15}.

4.8. Calculate the minimum number of raw PUF response bits needed to generate a 128-bit stable response given using a code-offset scheme based on a BCH decoder. It is assumed that the estimated error rate of the PUF raw output is 8% and the decoding error of the BCH scheme scheme should not exceed 10^{-6}.

4.9. A code-offset scheme, based on Hamming code C $(n, k, t) = (15, 11, 7)$ from Example 4.7, is used for reliable response generation for a PUF whose response length is 15 bits and maximum bit error rate is 5%.

Table 4.4 Oscillation frequencies of inverter chains of an RO PUF

Chain	R1	R2	R3	R4	R5
Frequency (MHz)	214	213	223	230	207

Table 4.5 List of reliable response generation schemes with their associated area overheads

Technique type	Area ($um2$)	Energy dissipation (pJ)	Bit error rate reduction
Code-offset scheme	15,130	13.2	6
Syndrome coding	16,320	15.1	7
Temporal majority voting	9750	6.4	5
Hardware redundancy	1120	4.5	3

(a) Explain whether or not the above reliable response generation scheme is suitable for cryptographic key generation.

(b) Compute a helper data vector for the following PUF response r = (000110111011101).

4.10. Consider a ring oscillator PUF that has five inverter chains whose respective frequencies are shown in Table 4.4.

Assume that the maximum variation in frequency due to noise is ± 4 MHz.

(a) Calculate the total number of challenge/response pairs for this PUF.

(b) IBS approach was applied to this PUF such that the number of bit flips caused by noise is reduced to zero. Recompute the number of challenge/response pairs for this reliability-enhanced PUF.

4.11 Table 4.5 includes a list of reliable response generation schemes with their associated area overheads, power consumption and the reduction they can achieve in the bit error rate of a PUF response.

It is assumed that the expected bit error rate of the PUF output is 9%. Study the specification of the following two designs carefully and propose a technique or a combination of techniques for reliable PUF response generation in each case

1. Cryptographic key generation block using a PUF for an extremely energy-constrained device.
2. An authentication tag for a commercial product with stringent area constraints. It is assumed, in this case, the tag reader is able to tolerate up to 4% of errors.

References

1. A. Maiti, P. Schaumont, The impact of aging on a physical unclonable function. IEEE Trans. Very Large Scale Integr. VLSI Syst. **22**, 1854–1864 (2014)
2. M.S. Mispan, B. Halak, M. Zwolinski, NBTI aging evaluation of PUF-based differential architectures, in *2016 IEEE 22nd International Symposium on On-Line Testing and Robust System Design (IOLTS)* (2016), pp. 103–108
3. D. Ganta, L. Nazhandali, Study of IC aging on ring oscillator physical unclonable functions, in *Fifteenth International Symposium on Quality Electronic Design* (2014), pp. 461–466
4. C.R. Chaudhuri, F. Amsaad, M. Niamat, Impact of temporal variations on the performance and reliability of configurable ring oscillator PUF, in *2016 IEEE National Aerospace and Electronics Conference (NAECON) and Ohio Innovation Summit (OIS)* (2016), pp. 458–463
5. B. Halak, A. Yakovlev, Statistical analysis of crosstalk-induced errors for on-chip interconnects. IET Comput. Digital Tech. **5**, 104–112 (2011)
6. G.I. Zebrev, A.M. Galimov, Compact modeling and simulation of heavy ion induced soft error rate in space environment: principles and validation. IEEE Trans. Nucl. Sci. 1–1 (2017)
7. V. Vargas, P. Ramos, V. Ray, C. Jalier, R. Stevens, B.D.D. Dinechin et al., Radiation experiments on a 28 nm single-chip many-core processor and SEU error-rate prediction. IEEE Trans. Nucl. Sci. **64**, 483–490 (2017)
8. C. Yanni, K.K. Parhi, Small area parallel Chien search architectures for long BCH codes. IEEE Trans. Very Large Scale Integr. VLSI Syst. **12**, 545–549 (2004)
9. S.Y. Wong, C. Chen, Q.M.J. Wu, Low power Chien search for BCH decoder using RT-level power management. IEEE Trans. Very Large Scale Integr. VLSI Syst. **19**, 338–341 (2011)
10. M. Bhargava, K. Mai, A high reliability PUF using hot carrier injection based response reinforcement, in presented at the *Proceedings of the 15th International Conference on Cryptographic Hardware and Embedded Systems*, Santa Barbara, CA (2013)
11. S. Satpathy, S. Mathew, V. Suresh, R. Krishnamurthy, Ultra-low energy security circuits for IoT applications, in *2016 IEEE 34th International Conference on Computer Design (ICCD)* (2016), pp. 682–685
12. M.D. Yu, S. Devadas, Secure and robust error correction for physical unclonable functions. IEEE Des. Test Comput. **27**, 48–65 (2010)
13. M. Cortez, A. Dargar, S. Hamdioui, G.J. Schrijen, Modeling SRAM start-up behavior for physical unclonable functions, in *2012 IEEE International Symposium on Defect and Fault Tolerance in VLSI and Nanotechnology Systems (DFT)* (2012), pp. 1–6
14. B.C. Paul, K. Kunhyuk, H. Kufluoglu, M.A. Alam, K. Roy, Impact of NBTI on the temporal performance degradation of digital circuits. IEEE Electron Device Lett. **26**, 560–562 (2005)
15. H.K.M.A. Alam, D. Varghese, S. Mahapatra, A comprehensive model for PMOS NBTI degradation: recent progress. Microelectron. Reliab. **47**, 853–862 (2007)
16. M. Bhargava, C. Cakir, K. Mai, Reliability enhancement of bi-stable PUFs in 65 nm bulk CMOS, in *2012 IEEE International Symposium on Hardware-Oriented Security and Trust* (2012), pp. 25–30
17. G. Pobegen, T. Aichinger, M. Nelhiebel, T. Grasser, Understanding temperature acceleration for NBTI, in *2011 International Electron Devices Meeting* (2011), pp. 27.3.1–27.3.4
18. X. Li, J. Qin, J.B. Bernstein, Compact modeling of MOSFET wearout mechanisms for circuit-reliability simulation. IEEE Trans. Device Mater. Reliab. **8**, 98–121 (2008)
19. G.E. Suh, S. Devadas, Physical unclonable functions for device authentication and secret key generation, in *2007 44th ACM/IEEE Design Automation Conference* (2007), pp. 9–14
20. M. Hofer, C. Boehm, An alternative to error correction for SRAM-like PUFs, in *Proceedings of Cryptographic Hardware and Embedded Systems, CHES 2010: 12th International Workshop*, Santa Barbara, USA, 17–20 August 2010, ed. by S. Mangard, F.-X. Standaert (Springer, Berlin, 2010), pp. 335–350
21. Y. Dodis, L. Reyzin, A. Smith, Fuzzy extractors: how to generate strong keys from biometrics and other noisy data, in *Proceedings of Advances in Cryptology - EUROCRYPT 2004:*

International Conference on the Theory and Applications of Cryptographic Techniques, Interlaken, Switzerland, 2–6 May 2004, ed. by C. Cachin, J.L. Camenisch (Springer, Berlin, 2004), pp. 523–540

22. F. Armknecht, R. Maes, A.-R. Sadeghi, B. Sunar, P. Tuyls, Memory leakage-resilient encryption based on physically unclonable functions, in *Proceedings of Advances in Cryptology – ASIACRYPT 2009: 15th International Conference on the Theory and Application of Cryptology and Information Security*, Tokyo, Japan, 6–10 December 2009, ed. by M. Matsui (Springer, Berlin, 2009), pp. 685–702

23. J. Delvaux, D. Gu, D. Schellekens, I. Verbauwhede, Helper data algorithms for PUF-based key generation: overview and analysis. IEEE Trans. Comput. Aided Des. Integr. Circuits Syst. **34**, 889–902 (2015)

24. Y. Lao, B. Yuan, C.H. Kim, K.K. Parhi, Reliable PUF-based local authentication with self-correction. IEEE Trans. Comput. Aided Des. Integr. Circuits Syst. **36**, 201–213 (2017)

25. W. Trappe, R. Howard, R.S. Moore, Low-energy security: limits and opportunities in the internet of things. IEEE Secur. Priv. **13**, 14–21 (2015)

26. R. Maes, A. Van Herrewege, I. Verbauwhede, PUFKY: a fully functional PUF-based cryptographic key generator, in *Proceedings of Cryptographic Hardware and Embedded Systems – CHES 2012: 14th International Workshop*, Leuven, Belgium, 9–12 September 2012, ed. by E. Prouff, P. Schaumont (Springer, Berlin, Heidelberg, 2012), pp. 302–319

27. B. Halak, Partial coding algorithm for area and energy efficient crosstalk avoidance codes implementation. IET Comput. Digital Tech. **8**, 97–107 (2014)

Security Attacks on Physically Unclonable Functions and Possible Countermeasures

5

5.1 Introduction

PUF technology promises to be the basis of new hardware-based security solutions that are highly resilient to physical attacks. The security of PUFs is due to the extreme difficulties in reproducing the identical implementation of an integrated circuit on multiple devices. This is particularly true for silicon-based designs, which cannot be cloned even by their original manufacturers.

However, the ongoing research in this field has shown that existing PUF implementations are vulnerable to a number of security threats including modelling attacks using machine learning algorithms, reverse engineering, side channel analysis, fault injections and helper data leakage [1–4]. This poses some interesting questions: *how secure PUFs really are, and how to reason about their security*.

The security of classic cryptographic primitives (e.g. ciphers, one-way functions, etc.) is normally assessed using the following two criteria:

(a) Their resilience to best known attacks.
(b) The inherent qualities of their design.

Let us take symmetric (shared key) ciphers as an example, in this case, there are a number of known attacks against these primitives, including but not limited to brute force key search, differential fault analysis and side channel attacks. In addition, a symmetric encryption scheme has a number of quality metrics that can be assessed such as the size of their key space and their avalanche effect. The latter is a measure of the average number of the output bits that change when on input bit changes.

© Springer International Publishing AG, part of Springer Nature 2018
B. Halak, *Physically Unclonable Functions*,
https://doi.org/10.1007/978-3-319-76804-5_5

The advanced encryption standard (AES) [5] is considered to be one of the most secure symmetric encryption algorithms due to its proven resilience to all known attacks. It also has a number of desirable security properties such as a strong avalanche effect, which means a significant number of its ciphertext output bits change if one bit changes in its plaintext input or key, this means it is very hard to find a statistical correlation between the ciphertext and the plaintext. AES algorithm has a massive key space, which makes it unfeasible to find the encryption key using brute force search.

The same evaluation approach can also be applied to assess the security of PUF designs, this means, one needs to have an understanding of all known attacks on PUFs, and develop a number of metrics and benchmarks which can be used to evaluate and test the security-related features of a PUF design.

This chapter aims to:

1. Introduce a number of quality metrics to evaluate the security of a PUF design.
2. Explain the principles of existing attacks on PUFs and their respective countermeasures.

It is hoped that this chapter will give the reader the necessary theoretical background and skills to understand PUF security attacks, evaluate the suitability of a PUF design with respect to these threats and develop appropriate countermeasures.

5.2   Chapter Overview

The organisation of this chapter is as follows: Sect. 5.3 discusses the design qualities one should consider when evaluating the security of a PUF design. In Sect. 5.4, we look in details into the metrics that quantify these qualities. Section 5.5 classifies the types of adversaries commonly encountered in the context of PUF security. Sections 5.6, 5.7 and 5.8 explain the principles of machine learning algorithms and how these can be employed to wage mathematical cloning attacks; this is followed by a detailed discussion of suitable countermeasures in Sect. 5.9. Sections 5.10 and 5.11 discuss existing side channel attacks and their related countermeasures, respectively. Section 5.12 explains the principles of physical attacks; followed by a discussion of possible defence mechanisms in Sect. 5.13. A comparative analysis of existing PUF attacks and associated countermeasures is presented in Sect. 5.14; this is followed by conclusions and lessoned learned in Sect. 5.15. A list of problems and exercises is provided in Sect. 5.16.

5.3   What Is a Secure PUF?

Since its conception more than a decade ago [6], the PUF technology has seen a dramatic increase in the number of new implementations as we have seen in Chap. 2.

The question is *how to choose a suitable PUF for a particular application*, in other words, what criteria should a designer be looking for in a PUF when constructing a secure system.

In principle, there are four fundamental properties, which allow for an objective comparison of the security of different PUF designs, namely:

- *Randomness* ensures the balance between 0 and 1 in the PUF responses, this makes it harder for an adversary to deduce information about the PUF by observing its output.
- *Physical Unclonability* ensures that an adversary is not able to create another physical instance of a PUF which has identical challenge/response behaviour.
- *Unpredictability* ensures that an adversary is not able to predict the PUF response to a new challenge, even if he had access to a large set of previous challenge/response pairs. This property can also be referred to as *Mathematical Unclonability*.
- *Uniqueness* ensures that each PUF instance has a unique behaviour, such that an adversary, who manages to wage a successful attack on a PUF instance, will not be able to predict the behaviour of other instances of the same PUF implemented on different devices.

The relative importance of these qualities may be different from one application to another. For example in key generation schemes, *Uniqueness* is the most important in order to ensure that different devices have distinct derived keys, whereas in authentication schemes, *unpredictability* is more significant in order to prevent the construction of a behavioural model by an adversary snooping on the communication between a PUF and an authenticating server. On the other hand, the use of PUF as a true random bit generator (TRNG) requires a design that has the maximum possible *randomness*.

In addition, a good understanding of the intended application is also important to help identify the most likely security threats. For example, in wireless sensor networks, devices may be deployed in the field without any physical protection, therefore, reverse engineering and side channel analysis are very likely risks, this may not be true for large data servers normally located in well protected locations.

From the above discussion, we can conclude that in order to answer the question '*how secure a PUF is*', we first need to understand their intended application, this can help to identify the most important design qualities to consider and the most likely threat model. In addition, we need to evaluate their robustness against known attacks. Figure 5.1 depicts a visual illustration of these three elements.

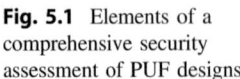

Fig. 5.1 Elements of a comprehensive security assessment of PUF designs

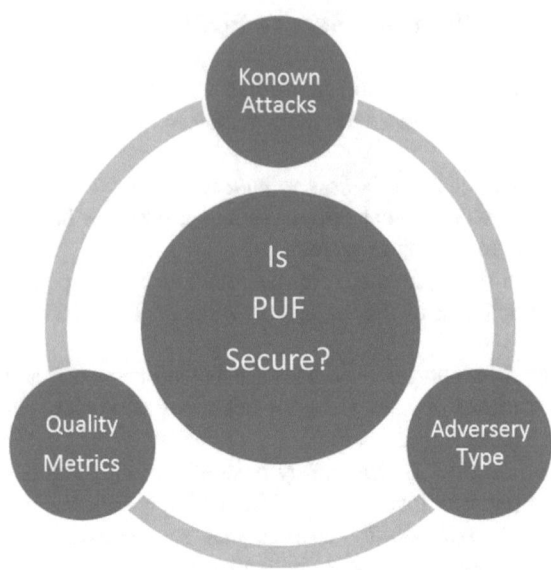

5.4   Security Evaluation Metrics for PUF

This section gives formal definitions to the security properties discussed above and presents the metrics which can be used to quantify these qualities.

5.4.1   Notation

We will use the following notations in the rest of this chapter. A PUF implementation is referred to using a letter I. Challenge and responses are referred to the letters C and R, respectively. Each PUF is considered to have n inputs and m outputs. Adversaries are referred to using the letter A.

5.4.2   Randomness

Consider a PUF device I which has a set of valid challenges C_V and their corresponding valid responses R_V.

The PUF I is said to have a random response if the probability of an output '1' is the same as that of an output '0'.

There are a number of approaches to evaluate *randomness* as follows:

(a) *Uniformity*

Uniformity estimates the percentage of '1's and '0's in a PUF response [7], it can be computed as follows:

$$Uniformity = \frac{1}{m * |R_v|} \sum_{r \in Rv} HW(r_i) \times 100\% \qquad (5.1)$$

where $|R_v|$, m are the total number of responses and their bit length, respectively. $HW(r_i)$ is the Hamming weight of the ith response.

A truly random variable will have an equal percentage of '1's and '0's, therefore a uniformity of 50%.

(b) *Statistical Tests*

The randomness of PUF responses can be estimated experimentally by applying statistical tests which are based on stochastic techniques. Examples of these test suits include NIST [8] and DIEHARD [9].

However, it might not be possible to rely on these approaches if there are not sufficient data, for example the case of memory-based PUF with small-sized challenge/response pair's space.

(c) *Entropy*

A more generic approach is the use of *entropy* as defined by Shannon [10, 11], it is commonly used in cryptographic applications to quantify how unpredictable a given discreet random variable is.

Let $X(r)$ be a random binary variable representing all responses. The entropy $(H(X))$ is computed as follows:

$$H(X) \triangleq - \sum_{r \in X} (p(r) \log_2 p(r)) \qquad (5.2)$$

where
$r \in \{0, 1\}$
$p(r)$ is the probability of a response (r).

In this context, one can use the *min-entropy*, introduced by Renyi [12], which is a more pessimistic notion of Shannon entropy, it is given as follows:

$$H_\infty(X) \triangleq - \log_2 \max_{r \in X} p(r) \qquad (5.3)$$

If X is a uniformly distributed discrete variable, then its min-entropy is equal to its Shannon entropy, for all other distributions, the min-entropy is upper bounded by Shannon entropy, i.e. $H_\infty(X) < H(X)$.

5.4.3 Physical Unclonability

Consider a PUF device I which has a set of valid challenges C_V and their corresponding valid responses R_V.

Let I_C be a physical clone of I, created by an adversary A. Assume I_C can generate a set of responses R_C for the same set of challenges C_V.

The PUF I is said to be physically unclonable if the average Hamming distance between the elements of R_V and their corresponding counterparts in R_C is significantly larger than the average intra-Hamming distance between the responses to the same challenge HD_{INTRA}, taken at different environmental/noise conditions.

This can be mathematically expressed using the following equation:

$$\frac{1}{|C_v|} \sum_{c \in C_V} d(rc, rv) \gg HD_{INTRA} \quad \forall \{A\} \tag{5.4}$$

where

rv, rc are the responses generated by the device and its clone respectively, for the same challenge c.

$|C_v|$ is the total number of valid challenges.

In other words, the difference between the responses generated by a clone should be easily distinguished from those created by an authentic device, and the difference between the two cannot be mistakenly attributed to the errors caused by noise or environmental variations.

By definition, all PUFs have to be *physically unclonable*, but recent advances in reverse engineering techniques have made physical cloning a possibility in some cases [3, 13].

There is no single metric that can be used to evaluate how physically unclonable a PUF is, but Eq. (5.4) can be used to examine how successful a physical cloning is.

5.4.4 Unpredictability (Mathematical Unclonability)

Consider a PUF device I that has a set of valid challenges C_V and their corresponding valid responses R_V. Consider an *Adversary A* who has access to a large number of challenges C_A and their associated responses R_A.

Let C_p denote a set of valid challenges whose corresponding responses are unknown to *Adversary A*.

Let rp denote the response predicted by *Adversary A to a challenge* ($c \in C_p$) based on his previous knowledge, i.e. $\{C_A, R_A\}$, and rv denotes the valid response of the same challenge.

The PUF I is said to be unpredictable if the average Hamming distance between rp and rv is significantly larger than the average intra-Hamming distance between the responses to the same challenge HD_{INTRA}, taken at different environmental/noise conditions.

This can be mathematically expressed using the following equation:

$$\frac{1}{|C_p|}\sum_{c \in Cp} d(rp, rv) \gg HD_{INTRA} \forall \{A\} \tag{5.5}$$

This means the average bit error in the predicted responses is significantly larger than that caused by noise, in other words, it is not possible for any adversary to construct an accurate mathematical model of the PUF even if he has access to large number of challenge/response pairs.

There are a number of approaches to evaluate the *Unpredictability* as follows:

(a) *Conditional Entropy*

Let $X(rp)$ be a random variable representing all unknown responses.

Let $Y(ra)$ be a random variable representing all responses known to the adversary.

The unpredictability of the PUF can be estimated using Shannon conditional entropy as follows:

$$H(X/Y) \triangleq - \sum_{rp \in X} \sum_{ra \in Y} (p(rp, ra) \log_2 p(rp|ra)) \tag{5.6}$$

And the minimum conditional entropy can be computed as follows:

$$\bar{H}_\infty(X|Y) \stackrel{\text{def}}{=} - \log_2 \left(\max_{\substack{r \in X \\ r' \in Y}} p(rp/ra) \right) \tag{5.7}$$

The above equations quantify the average and minimum number of bits of rp that cannot be predicted by an adversary even if he has knowledge of ra.

The use of the conditional entropy metric as described above may not be feasible in practice due to the large size of the challenge/response space. In addition, the computation of the conditional probability $p(rp/ra)$ is not straightforward, this is because both rp, ra are generated from the same PUF, so cannot be considered independent. The dependency between rp and ra can only be established if detailed knowledge of the physical implementation is known, and some simplifying assumptions are made. Examples of the use of this metric can be found in [14, 15].

(b) *Prediction Error of Machine Learning Algorithms*

In order to evaluate the *Unpredictability* of a PUF design, one can evaluate how 'learnable' the statistical relationship between the challenges and the responses is using machine learning algorithms [16]. To do this, one has to obtain a large number of /challenge/response pairs, some of these data are used in the 'training' stage to construct a software model of the PUF. The latter is then verified in the 'validation' stage using the remainder of the challenge/response pairs.

Once a model is created, the prediction error is computed by comparing the average Hamming distance between the predicted responses and those given by the model for the same set of challenges.

(c) *Hamming Distance Test*

This metric can be useful in developing a deeper understanding of correlations that may exist between challenges with a specific Hamming distance.

Let us consider a pair of challenges (c_j, c_i), with a Hamming distance $d(c_j, c_i) = t$, and corresponding responses (r_j, r_i). Let e represent the mismatch patterns between the challenge pair $e = c_j \oplus c_i$.

The Hamming distance test metric, denoted as $HDT(e, t)$, measures the output transition probability of a PUF for a given mismatch pattern e with Hamming weight t, that is $t = HW(e)$. It is computed as

$$HDT(t, e) = \Pr\big[(r_j \oplus r_i) = 1\big] \qquad (5.8)$$

A PUF is said to be unpredictable if $HDT(t, e) = 0.5$ for a given Hamming distance $d(c_j, c_i) = t$ and all its corresponding patterns [17]. This condition can be translated into two requirements:

(a) The mean value (μ) of the $HDT(t, e)$ should be 0.5
(b) The ratio between the standard deviation (σ) of the all computed $HDT(t, e)$ values and its mean should be close to zero

These two requirements can be mathematically expressed as follows:

$$\mu(HDT(t, e)) = \frac{1}{|C|} \sum_{c \in C} HDT(t, e, c) \approx 0.5 \qquad (5.9)$$

$$\frac{\sigma(HDT(t, e))}{\mu(HDT(t, e))} = \frac{\sqrt{\dfrac{\sum_{c \in C} (HDT(t, e, c) - \mu)^2}{|C| - 1}}}{\mu(HDT(t, e))} \approx 0 \qquad (5.10)$$

$|C|$ is the total number of challenges being considered for this test.

In most PUF designs, the responses of two challenges with a small Hamming distance have some dependencies. For example, in an SRAM PUF, wherein an address is used as a challenge, two closely related challenges may point to two memory cells with a close physical proximity. Therefore, the dimensions of these cells are likely to be less affected by intra-die process variations, hence their respective transistors threshold biases can be very similar, consequently their corresponding start-up values (i.e. responses) can be identical, which means the HDT metric in this case is likely to be 0.

Another example, wherein two challenges with small Hamming distance have dependencies, is the arbiter PUF. In this case, such challenges may result in highly similar choice of signal paths, which may lead to identical responses.

Therefore in such cases, designers are likely to have a more realistic evaluation of the *Unpredictability* of the PUF if they use $t = 1$, to compute the $HDT(e,t)$ metric.

5.4.5   Uniqueness

Consider a PUF device I that has a set of valid challenges C_V and their corresponding valid responses R_V.

A PUF is considered to be unique if it can be easily distinguished from other PUF with the same structure implemented on different chips in the same technology.

This metric is evaluated using the 'Inter-chip Hamming Distance'. If two chips, i and j ($i \neq j$), have n-bit responses, $R_i(n)$ and $R_j(n)$, respectively, for the challenge c, the average inter-chip HD among k chips is defined as [7]

$$HD_{INTER} = \frac{2}{k(k-1)} \sum_{i=1}^{k-1} \sum_{j=i+1}^{k} \frac{HD\big(R_i(n), R_j(n)\big)}{n} \times 100\% \qquad (5.11)$$

Ideally, the uniqueness of a PUF should be 50%.

5.4.6   An Exemplar PUF Security Evaluation

This section presents an exemplar case study of how to apply the above metrics to evaluate the security of a PUF design for a specific application.

Example 5.1 Table 5.1 shows the challenge/response pairs of a PUF design (n = 5, m = 1).

(a) Compute the uniformity of this design.
(b) Estimate Shannon Entropy and the min-Entropy of the PUF responses.
(c) Based on your answers in (a) and (b), explain whether this PUF can be used as truly random number generator.
(d) Use the Hamming distance metric $HDT(t,e)$ to evaluate the unpredictability of this PUF based on first five responses only and $t = 1$.
(e) Would you recommend this design for a device authentication protocol wherein a device incorporating this PUF is authenticated, if it can generate a valid response to a challenge presented by an authenticating server?
(f) Assuming the uniqueness of the above design is 49%, which application would you recommend this design for?

Table 5.1 Exemplar PUF challenge/response pairs

CRP	Challenge	Response
1	00000	0
2	00001	0
3	00010	1
4	00011	0
5	00100	1
6	00101	0
7	00110	0
8	00111	0
9	00100	1
10	01001	0
11	01010	1
12	01011	0
13	01100	1
14	01101	0
15	01110	0
16	01111	0
17	10000	1
18	10001	1
19	10010	0
20	10011	0
21	10100	0
22	10101	1
23	10110	1
24	10111	1
25	11000	1
26	11001	1
27	11010	0
28	11011	1
29	11100	0
30	11101	0
31	11110	1
32	11111	0

Solution:

(a) The uniformity can be computed using Eq. (5.1):

$$Uniformity = \frac{1}{k}\sum_{i=1}^{k} r_i \times 100\% = \frac{1}{32}\sum_{i=1}^{32}(0+0+1+\cdots) \times 100\% = 43\%$$

(b) Shannon Entropy can be computed using Eq. (5.2).

$$H(X) \triangleq -\sum_{x \in X} (p(x) \log_2 p(x)) = -(p(``1") \log_2 p(1) + p(``0") \log_2 p(``0"))$$
$$= -(0.4375 \log_2 (0.4375) + 0.57 \log_2 p(``0.5625"))$$
$$= 0.988$$

The min-entropy is given as

$$H_\infty(X) \triangleq -\log_2 \max_{x \in X} p(x) = -\log_2 \max\{p(``1"), p(0)\}$$
$$= -\log_2 \max\{0.4375, 0.5625\}$$
$$= -\log_2(0.5625)$$
$$= 0.83$$

(c) The uniformity of this PUF is 43%, which indicates its responses are more likely to be '0', in addition, the entropy of the responses is smaller than that of a truly random bit stream, which is 1,
The results from (a) and (b) both indicate that this design does not produce a random output, hence cannot be used as a random number generator.

(d) To evaluate $HDT(t = 1, e)$, we need to identify for each challenge c_j, the related challenges C_i such that $d(c_j, c_i) = 1, \forall c_i \in C_i$.
Then, we calculate the Hamming distance between each two respective responses $(r_j \oplus r_i)$.
The $HDT(1, e)$ for each challenge (r_j) is then computed as the average Hamming weight of $(r_j \oplus r_i)$ where $c_i \in C_i$. The results are summarised in Table 5.2.

Table 5.2 $HDT(t, e)$ evaluation results

(c_j, r_j)	c_i	r_j	$r_j \oplus r_i$	$HDT(1, e)$
(00000, 1)	00001	0	0	0.66
	00010	1	1	
	00100	1	1	
(00001, 0)	00000	0	0	0
	00011	0	0	
(00010, 1)	00000	0	1	1
	00011	0	1	
(00011, 0)	00001	0	0	0.5
	00010	1	1	
(00100, 1)	00000	0	1	0

Based on the results above, we can now compute the mean and standard deviation for the $HDT(1, e)$ using Eqs. (5.9) and (5.10) discussed earlier, in this case:

$$\mu(HDT(1, e)) = \frac{1}{|C|} \sum_{c \in C} HDT(1, e, c) = 0.43$$

$$\frac{\sigma(HDT(t, e))}{\mu(HDT(t, e))} = \frac{\sqrt{\frac{\sum_{c \in C} (HDT(t, e, c) - \mu)^2}{|C| - 1}}}{\mu(HDT(t, e))} = \frac{0.387}{0.43} = 0.9$$

Although the mean is close to the ideal value of 0.5, the standard deviation is too large, which indicates that there are overall strong correlation between the responses of challenges with Hamming distance of '1', especially in the case of the second, third and fifth challenges, where a response of a neighbouring challenge can be predicted with a 100% accuracy.

(e) The results from step (d) show that there is a dependency between the responses of closely related challenges. An adversary can exploit this to predict responses to unknown challenges using previously recorded responses of known challenges. Therefore, this design is not suitable for a device authentication scheme. This is because it is possible for an attacker to guess an authentic response to a presented challenge with high probability if he has recorded responses to a closely related challenge, which makes it possible for this adversary to authenticate fake devices.

(f) The uniqueness of this PUF is very close to the ideal value, which makes it a suitable candidate for a key generation scheme, as it allows generating distinct keys for different devices.

5.5 Types of Adversaries

In security applications, an adversary refers to a malicious party attempting to gain unauthorised access to services or resources, in order to steal information, take control or cause disruptions.

Stealing information means hacking into a system and collecting sensitive data (e.g. credit card details stored on a mobile device, patients' personal records located on a server in a medical centre, etc.).

Taking control is a step beyond the above, it includes compromising authentication mechanisms, modifying sensitive data, denying users access to resources and reporting false information (e.g. reporting false smart metre readings).

Causing disruption, in general terms, means overwhelming a system with a flood of messages to reduce its ability to function or to bring it down, a prime example of this is denial of service attacks.

The notion of the security of cryptographic systems can only be discussed with respect to a particular adversary; therefore it is important to understand the assumptions made about the capabilities of the attackers in terms of skills, physical access and or technical knowledge about the design.

In the context of PUF technology, there are three main types of attackers who are classified according their assumed knowledge and capabilities, as follows.

5.5.1 Eavesdropping (Snooping) Adversary

This type of attackers is an intruder capable of monitoring and recording information transmitted over a network, he takes advantage of unsecured communication channels. Public Wi-Fi networks is a prime example of the latter, wherein attackers can easily obtain a password and install a monitoring software on one of communicating nodes which can listen to the transmission and sniff out users names and passwords. The same approach can be used to obtain a large number of challenge/response pairs of a PUF. The latter are transmitted in the clear form (i.e. unencrypted), in standard PUF authentication schemes [3].

5.5.2 Black Box Adversaries

This adversary is assumed to be capable of having a physical access to the interface of the silicon chip incorporating the PUF, and can apply a polynomial number of challenges to the device and collect the corresponding responses. He can also run the hardware to execute a specific behaviour to extract some wanted information like timing, power consumption or electromagnetic leaks.

One scenario, where this type of attackers can be a concern, is the use of PUF technology for securing the Internet of things (IoT). The vast majority of IoT devices are deployed in the field with no physical protection such as industrial sensors applications and smart building control systems. In such cases, the adversary can easily obtain one of such devices and carry out various tests.

5.5.3 White Box Adversaries

In this case, the attacker is assumed to be able to gain knowledge of the layout and logic structure of the silicon chip incorporating the PUF. This adversary is assumed to be capable of all known semi-invasive and invasive attacks including, reverse engineering, fault injections and micro-probing.

This threat level can be a likely possibility due the multinational nature of chip production chains these days. In fact, the vast majority of silicon chips are fabricated and tested in untrusted factories and testing facilities, this makes it easier for adversaries to acquire information about the physical architecture and functionality of designed chips [18].

5.6   Mathematical Cloning Attacks

Now we have a better understanding of the type of adversaries, we are going to explain the principles of mathematical cloning. These attacks aim at building a numerical model of a physically unclonable function, if successful, the obtained software copy of the PUF will be able to accurately mimic the behaviour of the original device, typically with less than 1% error. The process of obtaining a software clone of a PUF is dependent on the type of machine learning algorithm adopted, however, this process has a common procedure summarised in Fig. 5.2. First, the adversary should obtain sufficient number of challenge/response pairs, the latter are divided into two data sets, the first is used for training and the second for cross-validation. In the second step, the adversary constructs a parametric functional model of the PUF (*FM*), which takes, as an input, the challenges applied to the PUF (C_i) and internal values of the fabrication-dependent physical parameters of the PUF circuits. The latter can sometimes be referred to as the feature vector ($\vec{v}$), which can be multidimensional and has real and rational values. This functional model can then be used to predict the corresponding response bits, i.e. $FM(C_i, \vec{v}) = R_i$. In the third step, a suitable machine learning algorithm is employed to generate a mathematical clone of the PUF based on the training data set and the parametric model $FM(C_i, \vec{v})$ constructed in the second step. In the fourth step, the constructed PUF model is validated using the test data set, by comparing the predicted response for each challenge with the actual response from the test data. If acceptable prediction accuracy is achieved, the process finishes, otherwise it is repeated until a satisfactory model is obtained. It is worth indicating here, that it is in principle possible to build a model of the PUF without the need of an internal parametric model of the PUF (i.e. Step 2), but this will prolong learning process and may produce less accurate results.

5.7   Mathematical Cloning Using Support Vector Machine

In the previous section, we covered the basics of machine modelling algorithms, this section focuses on the support vector machine approach [19], one of the most widely used machine learning algorithms. We will explain its mathematical principles and give examples on how these are employed in the learning process.

5.7.1   Mathematical Definition

Support vector machine (SVM) is a supervised machine learning approach capable of performing binary classification of data. Such classification is achieved using a linear or a nonlinear hyperplane, the latter refers to the separation function (a line, a surface, etc.). To implement this approach, each data item is first plotted as a point

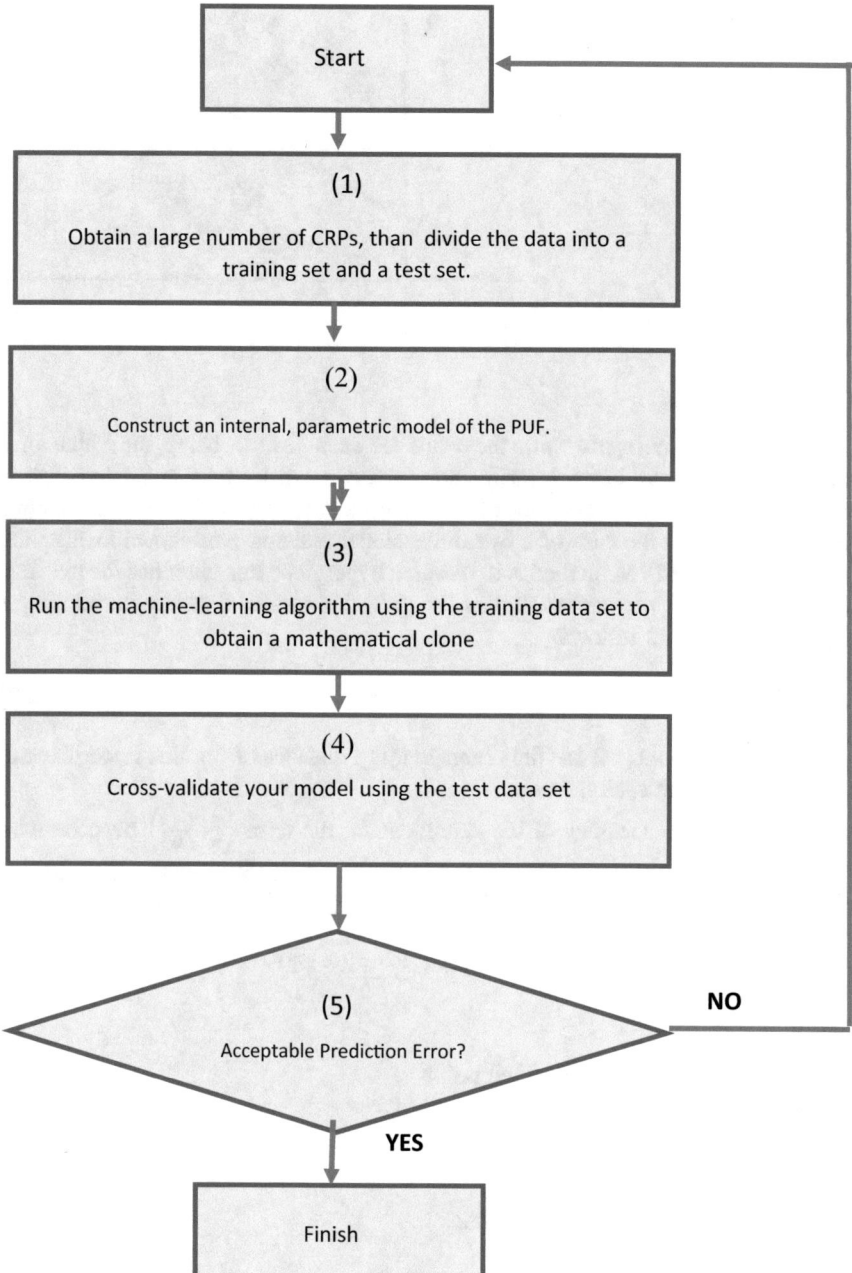

Fig. 5.2 A generic procedure for modelling physically unclonable functions using machine learning algorithms

Fig. 5.3 Support vector
machine hyperplane example

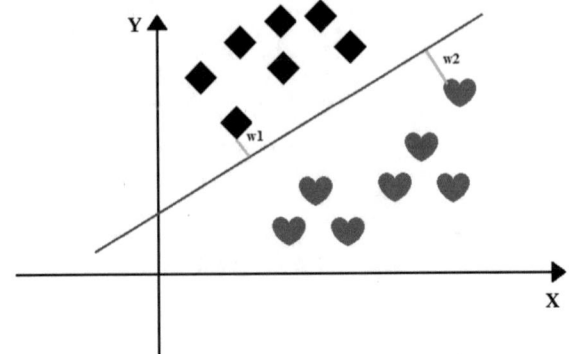

in multidimensional space with the value of each feature being the value of a particular coordinate. Then, classification is performed by finding the hyperplane that differentiates the two classes of data. Let us study the problem of classifying features vectors in the case of a two-dimensional real space as shown in Fig. 5.3.

The aim of the SVM method is to design a hyperplane that classifies the two data sets correctly. For the data shown in Fig. 5.3, this can be done by introducing a linear hyperplane as follows:

$$y = ax + b \tag{5.12}$$

Let us assume $w1, w2$ are the corresponding distances from this hyperplane to the closest point in each data set.

To increase the accuracy of the classification, the term $\left(\frac{1}{w1 + w2}\right)$ must be minimized, in addition, the number of misclassification must be minimised. These two conditions are equivalent to minimising the following cost function:

$$f(a, b) = iM + 1/(w1 + w2) \tag{5.13}$$

where

M　is the number of misclassified points.
i　is a positive integer.

5.7.2　Basic Working Principles

In this section, we are going to learn the basic working principles of SVM using illustrative examples.

Fig. 5.4 Three possible
hyperplanes for data
separation

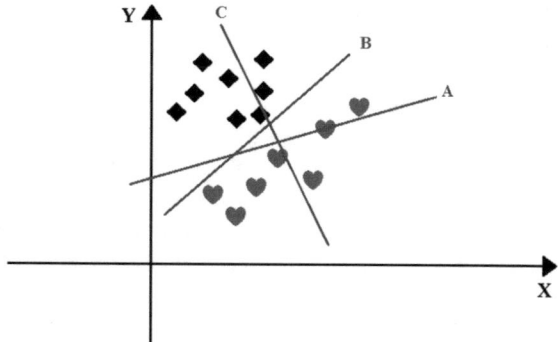

Example 5.2 In Fig. 5.4, there are two sets of data (diamonds and hearts); in
addition, there are three possible lines which can be used to separate these sets.
Study the data carefully and specify which of these lines is the best option?

In this case, we use a rule of thumb to identify the right hyperplane, which
intuitively states 'the better solution achieves better data segregation'. In this scenario, solution 'B' is certainly the best option, as it can achieve a clear separation
between the two sets.

The above rule of thumbs helps to reduce the number of misclassifications.

Example 5.3 Figure 5.5 shows three possible separating lines, which of these is the
best option?

In this case, applying the previous rule of thumb does not help as all lines can
segregate the data classes perfectly. In this case, we need to use a second rule of
thumb, which states 'the best solution maximises the distances between the nearest
data points (from either class) and the separating line'.

In this case, line B has larger distances from the points of both data sets compared to both A and C, therefore it is the best option.

This rule of thumb helps to allow for extra margin for error, which can help
reduce the number of misclassifications.

Fig. 5.5 Three possible
hyperplanes for data
separation

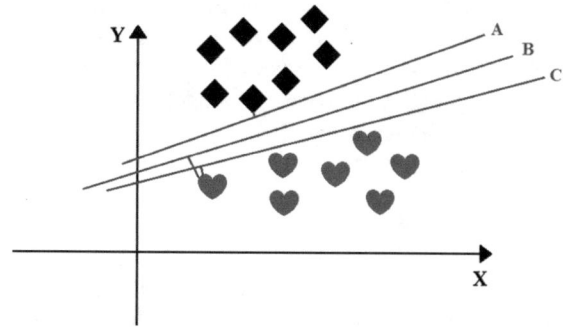

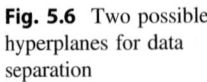 **Fig. 5.6** Two possible hyperplanes for data separation

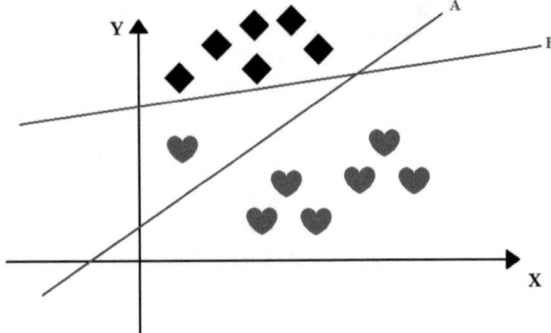

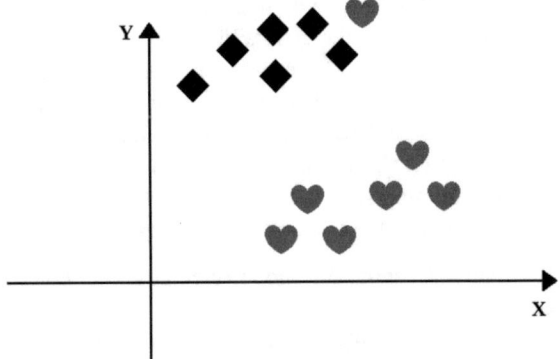 **Fig. 5.7** Two data classes with an outlier

Example 5.4 Figure 5.6 shows two possible separating lines which segregate two sets of data, which of these is the best option?

SVM selects the solution that classifies the classes accurately over those that maximise the minimum distances to the data points. Here, line A has a classification error and B classifies all points correctly. Therefore, we choose B.

Example 5.5 How to separate the two classes of data in Fig. 5.7?

In this case, it is not possible to find a solution using a straight line, as one of the heart points is located in a very close proximity to the diamonds. Such points are typically referred to as outliers. The latter are normally ignored by SVM during the learning process, otherwise a model cannot be extracted. In this example, the solution is shown in Fig. 5.8.

5.7.3   Kernels

In the previous examples, we have seen that data can be easily separated by a linear hyperplane; however, this may not be always the case. This section introduces the

Fig. 5.8 Data segregation in
the presence of an outliner

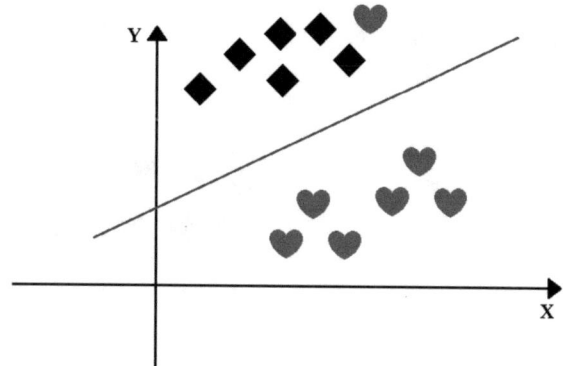

concept of *Kernels*, it also explains how these can help solve more complex classification problems.

Example 5.6 How to separate the two classes of data in Fig. 5.9?

This problem is typically resolved by SVM by introducing additional features.

First we note from Fig. 5.9, the hearts are closer to the origin of x and y axes than the diamonds are, based on this observation, we will add a new feature $(z = x^2 + y^2)$. Figure 5.10 shows a plot of the data points using axis x and z.

The addition of this new feature made it very easy to separate these two classes of data. This is typically referred to as the *kernel trick*. The latter consists of taking low dimensional input space and transform it to a higher dimensional space, these functions are called kernels, they are useful in nonlinear separation problems. The separation line in original input space looks like a circle (Fig. 5.11).

Fig. 5.9 Complex data
segregation problems

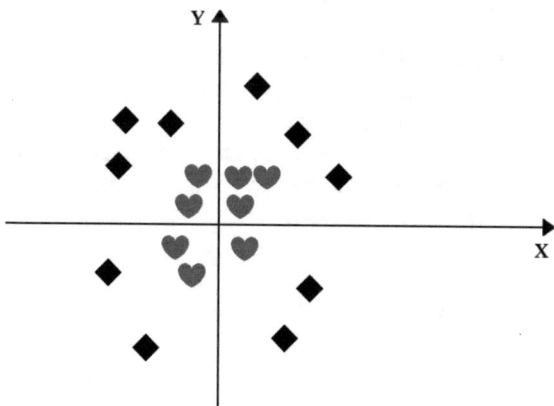

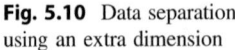

Fig. 5.10 Data separation using an extra dimension

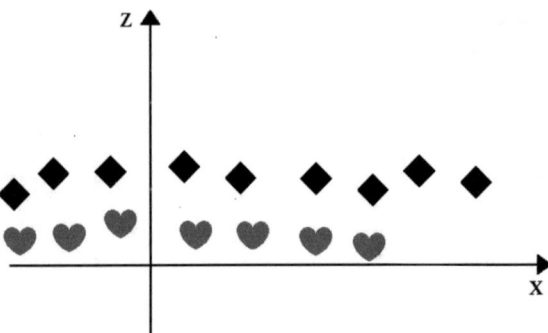

Fig. 5.11 Separating complex data using the kernel trick

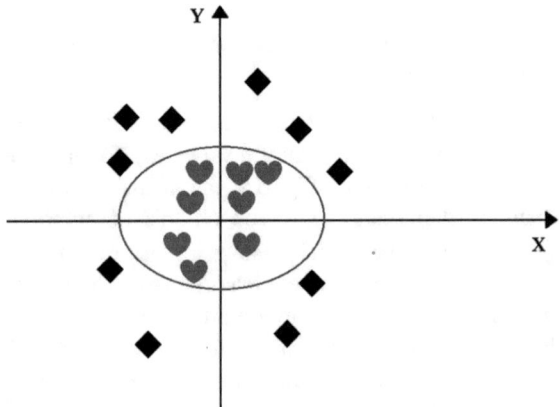

5.7.4   How to Set the SVM Algorithm Parameters

SVM functions are available in most mathematical simulation tools (e.g MATLAB) as library functions; it is important to understand how to define the parameters of these models in order to obtain accurate results. There are three parameters, which need to be set when we create an SVM classification object, namely: *kernel, gamma* and *C*. Table 5.3 summarises some available options for each of these parameters and their impact on the model.

5.7.5   Cross-Validation

Cross-validation techniques help to evaluate the accuracy of the prediction model. It consists of the following steps:

1. Divide the data into training set and test set
2. Build the prediction model using the training set

Table 5.3 Support vector machine parameters options

Parameters	Definition	Exemplar values	Impact on the model
Kernel	It defines the type of the complexity model created by SVM	Linear rbf poly	Linear are relevant to simple classification problems. 'rbf' and 'poly' are useful for nonlinear problems
Gamma	Kernel coefficient for 'rbf', 'poly' and 'sigmoid'. It define the accuracy of the precision model	0, 10, 100,…	If gamma is given a high value, the algorithm will try to create an exact fit as per training data set, this may lead to a generalisation error and cause an overfitting problem
C	Penalty parameter	0, 1, 1000	C controls the trade-off between smooth decision boundary and classifying the training points correctly

3. Estimate the prediction errors by comparing the real outputs from the test set and the predicted outputs from the model
4. Repeat the procedure k times, each time with a new training/test set

This is typically referred to as k-fold cross-validation.

5.7.6 Case Study: Modelling Arbiter PUF Using Support Vector Machine

Now, we have a better understanding of how SVM works, we are going to learn how to use it to model the behaviour of an arbiter PUF. A MATLAB toolbox is used to perform the analysis.

An arbiter PUF consists of k stages; each stage is composed of two 2-to-1 multiplexers as shown in Fig. 5.12. A rising pulse at the input propagates through two nominally identical delay paths. The paths for the input pulse are controlled by the switching elements, which are set by the bits of the challenge as shown in Fig. 5.13. For c = 0, the paths go straight through, while for c = 1 they are crossed. Because of manufacturing variations, however, there is a delay difference Δt between the paths. An arbiter at the end generates a response '0' or '1' depending on the difference in arrival times. In our simulation, a set–reset (SR) latch has been used as the arbiter block. We will now explain how to implement the modelling procedure described previously in Fig. 5.2.

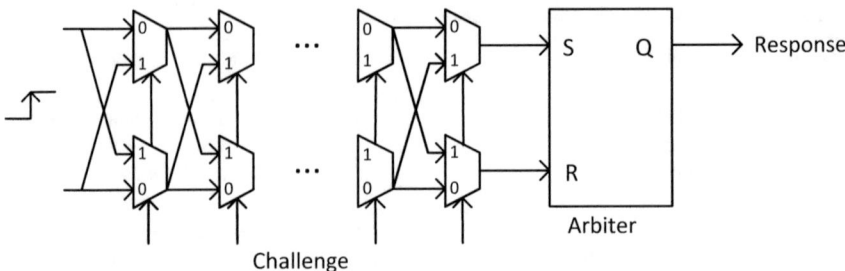

Fig. 5.12 A generic structure of an arbiter PUF

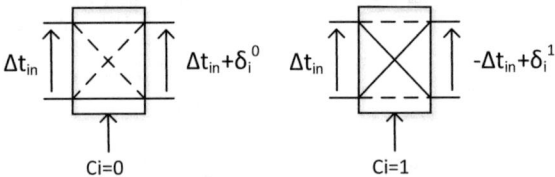

Fig. 5.13 The working principle of the multiplexer

Step 1: Obtain a large number of CRPs

A 16-bit arbiter PUF has been implemented in a 65-nm technology node and simulated using the BSIM4 (V4.5) transistor model with a nominal supply voltage 1.2 V and a room temperature 25 °C. Intrinsic variations such as effective length, effective width, oxide thickness and threshold voltage are modelled in Monte Carlo simulations using the built-in fabrication standard statistical variation (3σ variations) in the technology design kit. A total of 15000 CRPs have been generated for later analysis.

Step 2: Construct an internal, parametric model of the PUF.

The functionality of the arbiter PUF can be described by an additive linear model [6]. The total delays of both paths are modelled as the sum of the delays in each stage (switch component) depending on the challenge C (c1, c2...ck). The final delay difference Δt between the two paths in a k-bit arbiter PUF can be expressed as

$$\Delta t = \vec{D}^T \vec{v} \tag{5.14}$$

where parameter $\vec{D}$ is the delay-determined vector and $\vec{v}$ is the feature vector. Both parameters are functions of the applied k-bit challenge with dimension k + 1. As described in [20], we denote $\delta_i^{1/0}$ as the delay in stage i for the crossed (1) and uncrossed (0) paths, respectively. Hence, δ_i^1 is the delay of stage i when ci = 1, and δ_i^0 is the delay of stage i when ci = 0. Then

$$\vec{D}^T = \left(D^1, D^2 \ldots, D^K, D^{K+1}\right)^T \tag{5.15}$$

where

$$D^1 = \frac{\delta_1^1 + \delta_1^0}{2}$$
$$D^i = \frac{\delta_{i-1}^1 + \delta_{i-1}^0 + \delta_i^1 + \delta_i^0}{2}, \quad i = 2 \ldots k.$$
$$D^{k+1} = \frac{\delta_k^1 + \delta_k^0}{2}$$

In addition, we can write the feature vector $\vec{v}$ as follows:

$$\overrightarrow{v(C)} = \left(v^1(C), v^2(C) \ldots, v^K(C), 1\right)^T \tag{5.16}$$

where

$$v^1(C) = \prod_{i=j}^{k} (1 - 2C_i), \quad j = 2 \ldots k.$$

From Eq. 5.14, the vector $\vec{D}^T$ encodes the delay in each stage of the arbiter PUF and via $\vec{D}^T \vec{v} = 0$ we can determine the separating hyperplane in the space of all feature vectors $\vec{v}$. The delay difference, Δt, is the inner product of $\vec{D}^T$ and $\vec{v}$. If $\Delta t > 0$, the response bit is '1', otherwise, the response bit is '0'. Determination of this hyperplane allows building a mathematical clone of the arbiter PUF.

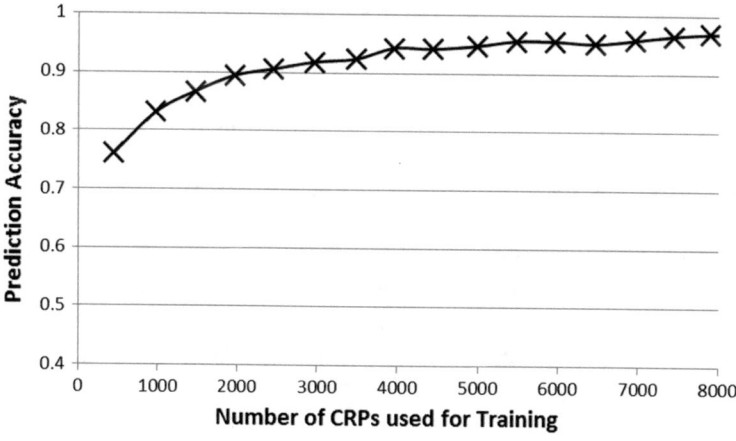

Fig. 5.14 Prediction accuracy of SVM model versus number of CRPs

Step 3: Running support vector machine algorithm using a training CRPs set

The model generated in Step 2 is used as an initial classification object; the algorithm is run multiple times to identify the best kernel which was found to be a linear kernel. Once the best kernel is determined, we incrementally increased the amount of training data, then extracted the associated mathematical models.

Step 4: Cross-validate your model using the test set

We performed a 5-fold cross-validation in order to enhance the accuracy of the derived model, 8000 training CRPs were sufficient in this case to construct a mathematical clone able to predict the responses with less than 1% error. Figure 5.14 shows the machine learning results. Readers are referred to Appendix C for an exemplar MATLAB script of the SVM attack.

5.8   Mathematical Cloning Using Artificial Neural Networks (ANN)

5.8.1   Mathematical Definition

An artificial neural network is a computational model that consists of a large collection of connected items referred to as artificial neurons [21]. The connections between these neurons carry activation signals with different strengths. If the combined strength of the incoming activation signals at the inputs of a neuron is sufficiently strong, the neuron will be activated and the signal travel through it to other connected neurons. The network has a layered structure. An ANN is typically defined by three types of parameters:

1. The interconnection pattern between the different layers of neurons.
2. The weights of the interconnections, which are updated during the learning process.
3. The activation function that converts a neuron's weighted inputs to its output.

Mathematically, a neuron's network activation function $f(x)$ is defined as a composition of other functions $g(x)$, which can further be defined as a composition of other functions. This can be conveniently represented as a network structure, with arrows depicting the dependencies between variables. A widely used type of composition is the *nonlinear weighted sum*, where

$$f(x) = K \sum_i w_i g_i(x) \tag{5.17}$$

K is a predefined function such as sigmoid.

An important consideration when choosing the activation function is to ensure it can provide a smooth transition as input values change, i.e. a small change in input produces a small change in output.

ANNs support a number of learning approaches, but supervised learning is the most relevant in the context of PUF modeling; this is assuming sufficient input/output date of the PUF are available.

The supervised learning approach aims to find a function $f : X \rightarrow Y$ from an allowed class of functions that minimise the prediction error. The latter is evaluated using a cost function (e.g. the mean square error).

There are many algorithms which can be used for training the neural networks for minimising this cost function, most notably the backpropagation algorithm [22].

5.8.2   Case Study: Modelling TCO PUF Using Neural Networks

This section gives an example of how to use ANN to model the behaviour of a PUF.

A typical structure of ANN is a feedforward network which can be constructed as single-layer perceptron (SLP) or multilayer perceptron (MLP). MLP has been considered in this study as it can solve a nonlinear problem. The MLP network consists of three layers of nodes which include an input layer, a hidden layer and an output layer. Except for the input layer, in each neuron, all input vector values are weighted, summed, biased and applied to an activation function to generate an output. A tan-sigmoid and linear transfer functions have been used for hidden and output layers, respectively. Using 32 neurons in the hidden layer is found to be an optimum setting. During training, an error resulting from the difference between a predicted and observed value is propagated back through the network and the weight and bias of the neurons are adjusted and updated. The training process stops when the prediction error reaches a predefined value or a predetermined number of epochs are completed. Based on our experiments, resilient backpropagation has been chosen as the best training algorithm considering the prediction accuracy and fast convergence time.

The above ANN has been implemented in MATLAB. In this case, we consider a 32-bit TCO PUF [23]. A set of 20000 CRPs are obtained by simulating the TCO PUF using 65 nm CMOS technology. 18000 CRPs were used for training and 2000 CRPs were utilised for testing. The results are shown in Fig. 5.15. Readers are referred to Appendix C for an exemplar MATLAB script of a ANN modelling attack.

5.9   Countermeasures of Mathematical Cloning Attacks

Defence mechanisms against modelling attacks can be classified into two types as follows:

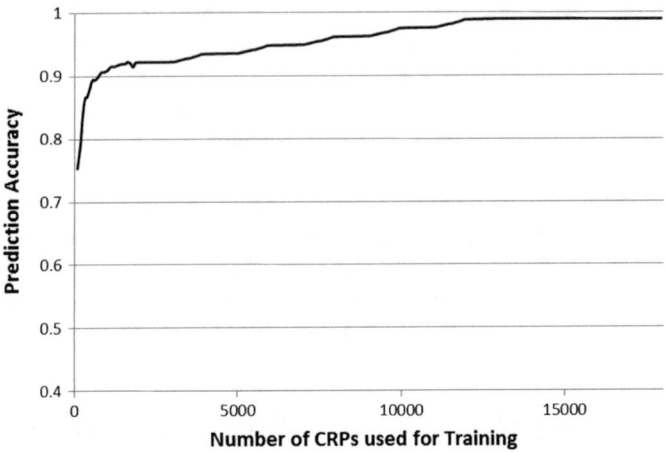

Fig. 5.15 Prediction accuracy of ANN model versus number of CRPs for a TCO PUF

(a) Prevention techniques that aim to stop an adversary from collecting enough challenge/response pairs to build a mathematical model of the PUF.
(b) Mitigation methods that aim to enhance the resilience of PUF designs against mathematical cloning using additional cryptographic primitives (e.g. Hash functions, encryption cores, etc.).

In this section, we are going to discuss a number of examples of each of the above categories. Before we explain these methods, we need to introduce the concept of multibit response PUFs. The latter refers to a PUF design that has an n-bit challenge and an m-bit-response, where $m > 1$. Some PUF designs can be easily configured to be multibits. For example, an SRAM PUF can generate a multibit response if the read/write circuitry of the memory block is designed to read multiple SRAM cells simultaneously. On the other hand, delay-based PUFs have typically single-bit outputs. In this case, multibit responses can be obtained using a number of single-bit output PUFs connected in parallel, such that the same challenge is applied to all of these simultaneously. A more cost-efficient approach is to apply a sequence of challenges on a single-bit output to generate the m-bit response vector, this approach can be further optimised by using a single challenge as a seed for a pseudorandom generator that outputs m corresponding challenges, the latter are applied in serial to the single-bit response PUF.

It should be noted that some of the techniques explained below implicitly assume that the PUF has a multibit response architecture.

5.9.1 Minimum Readout Time (MTR) Approach

This preventative countermeasure aims to severely restrict the ability of an adversary to collect sufficient number of challenge/response pairs that allow him to build a mathematical clone.

The essence of this approach is to significantly increase the latency of the PUF circuit such that the minimum time needed to retrieve enough CRPs to perform modelling attacks is impractically long.

One way to implement this technique is to include redundant logic at the output of the PUF circuit as shown in Fig. 5.16.

This technique may protect the PUF from snooping attacks wherein the adversary does not have access to the physical device, it can also be effective against black box attacks wherein the adversary only has access to the primary inputs/outputs of the silicon chip embedding the PUF.

However, this approach does not provide any protection against white box attacks, wherein it is possible to probe directly the internal signals inside the chip and bypass the delay block.

5.9.2 Partial Masking of Challenges/Response Bits

This is another preventative approach, its core principle is to partially mask the challenges (or responses) applied to (or generated from) the PUF. This partial masking approach makes it harder for machine learning algorithms to build an accurate model of the PUF.

One example of partial masking of multibit responses is shown [24] in the context of PUF authentication protocols. In this example, only a subset of PUF response strings is sent to the verifier during the authentication process, the verifying server uses a substring matching technique to validate the responses.

Another approach has been implemented in [25], wherein a partial challenge is sent by the verifier to the prover device embedding the PUF. The received short challenge is subsequently padded with a random pattern generated by a random number generator to make up a full-length challenge before being applied to the PUF; the server uses a challenge recovery mechanism to generate an emulated response to compare with the received response.

Both of the above approaches might increase the authentication time as well as consume more computing resources on the verifier side. One might argue, however, this may not be not a big concern since the verifier has always been assumed to be rich in resources.

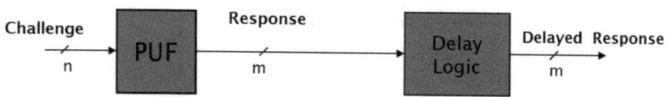

Fig. 5.16 Cloning prevention using minimum readout time

5.9.3 Partial Utilisation of Challenges/Response Pairs

This preventative approach was presented in [26], its principles are as follows: the challenges are divided into two subsets: valid and invalid, the former are called the secret challenges or (the s-challenges), the number of valid challenges in this set should not be sufficient to allow building a mathematical model. If we consider the example from Fig. 5.15, the s-challenges set in this case should contain only around 200 vectors to prevent building a model with accuracy more than 80%.

The PUF operates normally for the s-challenges; but if a challenge that is not in this subset is used, the PUF will not display an authentic, in other words, random data will be outputted as a response.

A possible implementation of this technique is shown in Fig. 5.17, wherein a challenge validation logic is used to check whether or not the applied challenge is part of the s-challenges, if yes then the challenge is applied to the PUF, otherwise it will be replaced by a random bit string generated locally by a random number generator.

5.9.4 Controlled PUFs

This is a mitigation technique wherein hash functions are used at the input/output of the PUF, which means an adversary may not be able to have a direct access to the challenge/response pairs, which in turn makes mathematical cloning of the PUF infeasible [27]. Although the use of hash function is a very effective approach, it incurs significant increase in chip area and energy costs, which may not be affordable in some applications such as low end IoT devices [28].

All of the above techniques can protect the PUF from a modelling attack by a snooping or black box adversary but do not provide any protection against white box adversaries.

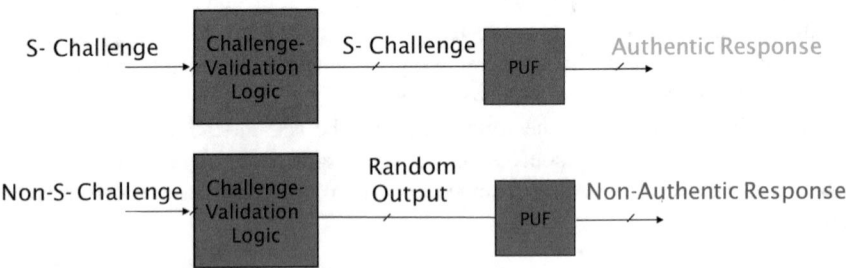

Fig. 5.17 Cloning prevention using partial utilization of CRPs

5.9.5 Permutation-Based Obfuscation

The notion of permutation consists of rearranging the order of the elements in set. This concept was employed to design a number of historical ciphers such as Skytale form the Greeks time wherein the encryption is done using a device which consists of a rod and a polygon base. To encrypt, one would write the message horizontally on a piece of paper wrapped around the device. When the paper is unwound, the letters on it would appear randomly written with no meaning as shown in Fig. 5.18. Only a legitimate receiver who has an identical device can decrypt the cipher.

Permutation techniques are also used in modern ciphers such as AES and triple-DES [5] to dissipate the redundancy of plaintext by spreading it all out over the cipher text, which makes it harder to establish a statistical correlation between the two. The same concept can be applied in the context of PUF in order to obscure the relationship between the challenges and their associated responses, hence make machine learning harder or infeasible [20].

One way to implement this approach is to permute the challenges as shown in Fig. 5.19, in this case the permutation block is placed before the PUF circuit. Each challenge is then replaced by another before being applied to the inputs of the PUF. This approach can be implemented on both single and multibit response designs.

The mechanism of permutation is as follows: each n-bit challenge is divided into k segments of length l wherein $l = \frac{n}{k}$. These segments are then permuted to generate a new challenge, as shown in Fig. 5.20.

Fig. 5.18 The skytale permutation cipher

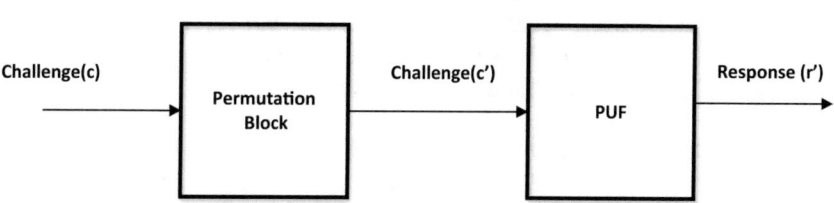

Fig. 5.19 Architecture of permuted challenge PUFs

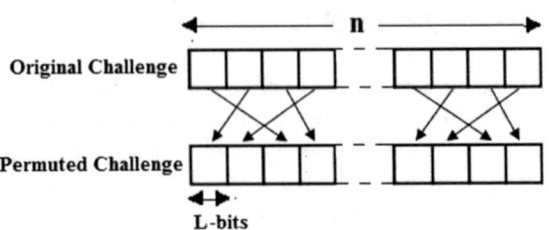

Fig. 5.20 Conceptual illustration of PUF challenge permutation

In principle, the larger the number of segments k(or the smaller the number of bit in each segments) the better results that can be achieved, as this will increase the number of possible permutation $k!$ which makes it harder for an adversary to find the original order of bits. The optimum way of permutation can be found using an iterative algorithm which searches for a solution to maximise the unpredictability metric as described in Sect. 5.4.2.

Response permutation is only possible for multibit response PUFs, but it may not be very useful on its own as it only changes the order of the response bits, so multibit machine learning algorithms can still build a mathematical model for each response bit.

Example 5.7 A PUF design has the CRPs' behaviour described in Table 5.4. For this design, two challenge permutation approaches are applied as follows:

(a) $1 \rightarrow 5, 2 \rightarrow 1, 3 \rightarrow 2, 4 \rightarrow 3, 5 \rightarrow 4$
(b) $1 \rightarrow 5, 2 \rightarrow 4, 3 \rightarrow 1, 4 \rightarrow 2, 5 \rightarrow 3$

The resulting challenge/response pairs in each case are shown in Table 5.4.

(1) Use the Hamming distance metric HDT(t,e) to evaluate the unpredictability of the original PUF using t = 1.
(2) Which of the above two permutation produces a less predictable design?

Solution:

(1) To evaluate $HDT(1, e)$, we need to identify for each challenge c_j, the related challenges c_i such that $d(c_j, c_i) = 1$, then calculate the Hamming distance between their respective responses $(r_j \oplus r_i)$. The $HDT(1, e)$ for each challenge is then computed as the average Hamming weight of $(r_j \oplus r_i)$ for each challenge. The results are summarised in Table 5.5.

Based on the results in Table 5.5, we can now compute the mean and standard deviation percentage for the $HDT(1, e)$ using Eqs. (5.9) and (5.10) as follows.

Table 5.4 PUF original and permuted responses

CRP	Challenge	Response	Permutation A	Permutation B
1	000	0	0	1
2	001	0	1	0
3	110	1	0	1
4	011	0	1	0
5	111	1	0	0

$$\mu(HDT(1,e)) = \frac{1}{|C|} \sum_{c \in C} HDT(1,e,c) = 0.2$$

$$\frac{\sigma(HDT(t,e))}{\mu(HDT(t,e))} = \frac{\sqrt{\frac{\sum_{c \in C}(HDT(t,e,c)-\mu)^2}{|C|-1}}}{\mu(HDT(t,e))} = \frac{0.245}{0.2} = 1.224$$

The results show both measures are far from their ideal values, which indicate the original design is predictable.

(2) To find the best permutation approach, we repeated the same computation as above for both the cases of permutation as summarised in Tables 5.6 and 5.7.

Table 5.5 $HDT(t,e)$ evaluation results

(c_j, r_j)	(c_i, r_i)	$r_j \oplus r_i$	$HDT(1,e)$
(000, 0)	(001, 0)	0	0
(001, 0)	(000,0)	0	0
	(011,0)	0	
(110,1)	(111,1)	0	0
(011,0)	(001,0)	0	0.5
	(111,1)	1	
(111,1)	(011,0)	0	0.5
	(110,1)	1	

Table 5.6 $HDT(t,e)$ evaluation results for Permutation A

(c_j, r_j)	(c_i, r_i)	$r_j \oplus r_i$	$HDT(1,e)$
(000, 0)	(001, 1)	1	1
(001, 1)	(000,0)	1	0.5
	(011,1)	0	
(110,0)	(111,0)	0	0
(011,1)	(001,1)	0	0.5
	(111,0)	1	
(111,0)	(011,1)	1	0.5
	(110,0)	0	

Table 5.7 $HDT(t,e)$ evaluation results for Permutation B

(c_j, r_j)	(c_i, r_i)	$r_j \oplus r_i$	$HDT(1, e)$
(000, 1)	(001, 0)	1	1
(001, 0)	(000,1)	1	1
	(011,1)	1	
(110,1)	(111,0)	1	0
(011,0)	(001,0)	0	0.5
	(111,0)	0	
(111,0)	(011,0)	0	0.5
	(110,1)	1	

Based on the results in Table 5.6, we can now compute the mean and standard deviation percentage for the $HDT(1, e)$ Eqs. (5.9) and (5.10) as follows:

$$\mu(HDT(1,e)) = \frac{1}{|C|} \sum_{c \in C} HDT(1, e, c) = 0.5$$

$$\frac{\sigma(HDT(t,e))}{\mu(HDT(t,e))} = \frac{\sqrt{\frac{\sum_{c \in C}(HDT(t,e,c)-\mu)^2}{|C|-1}}}{\mu(HDT(t,e))} = \frac{0.316}{0.5} = 0.63$$

Based on the results from Table 5.7, we can now compute the mean and standard deviation percentage for the $HDT(1, e)$ Eqs. (5.9) and (5.10) as follows:

$$\mu(HDT(1,e)) = \frac{1}{|C|} \sum_{c \in C} HDT(1, e, c) = 0.6$$

$$\frac{\sigma(HDT(t,e))}{\mu(HDT(t,e))} = \frac{\sqrt{\frac{\sum_{c \in C}(HDT(t,e,c)-\mu)^2}{|C|-1}}}{\mu(HDT(t,e))} = \frac{0.37}{0.6} = 0.63$$

The results show that permutation (A) improves the unpredictability of the design while permutation (B) gives a very similar profile to the original PUF behaviour, therefore, permutation (A) is the best option in this case.

5.9.6 Substitution-Based Obfuscation

The notion of substitution consists of replacing the elements in a set with new elements from a different set according to a fixed system. The main difference between substitution and permutation is that in the latter, the elements of the original set are re-ordered but left unchanged whereas substitution will replace the elements themselves with new set of elements. For example, let us assume we have a message which contains four digits $M = 1234$, a permutation function may reorder the digits to produce a permuted message as $Mp = 2413$, whereas a substitution function may replace the digits with letters and produce a substituted set $Ms = abcd$

Substitution techniques have been used in a number of historical ciphers such as Caesar shift cipher, there are also a fundamental part of modern block ciphers such as AES and triple-DES [5].

Substitution can be used in the case of PUF to mask the statistical relationship between the challenges and their corresponding responses.

There are two ways to implement this technique. The first method is based on substituting the responses, and the second is based on substituting the challenges. In both the cases, the substitution function should be chosen to be highly nonlinear such as the S-box used in the AES cipher [5].

Challenge substitution approach can be applied to single-bit and multibit response PUFs, while response substitution is only applicable to multibit designs,

A word of caution here: the use of the response substitution can increase the bit error rate at the output of the PUF. Consider, for example, the case of a 4-bit response PUF that has a substitution function at its output, which performs a one-to-one mapping of 4-bit data strings.

Let us assume in this example the string A = (0011) is substituted with C = (1111), and the string B = (0001) is substituted with D = (0000).

Now assume the challenge corresponding to response (A) was applied to the PUF, but environment variations causes a 1-bit flip in the least significant bit of the generated response, this effectively means that the PUF generates response (B) instead. This translates to a 25% error rate at the output of the PUF. Now if B is then applied to the substitution block, the latter will produce (C). If there was no bit flip, we would expect to get response (D) at the output of the substitution block, which means the error rate at the output of the substitution block is 100%.

5.9.7   Other Obfuscation Methods

The principles of challenges/response obfuscation can be implemented using a combination of substitution and permutation or with other security primitives such as block ciphers. In the following section, we will explore the effectiveness of the above approaches in increasing the resilience of PUF designs against machine learning algorithms.

5.9.8   Case Study: Evaluation of Mathematical Cloning Countermeasures

For this case study, we have implemented an arbiter PUF which has 16-bit input challenges and a single-bit output response in a 65-nm technology node. The design has been simulated using the BSIM4 (V4.5) transistor model with a nominal supply voltage 1.2 V and at a room temperature of 25 °C. Intrinsic variations such as effective length, effective width, oxide thickness and threshold voltage are modelled in Monte Carlo simulations using the built-in fabrication standard statistical variation in the technology design kit. The following schemes have also been implemented:

(a) **Permutation-Based Challenge Obfuscation**

This design uses the generic architecture shown in Fig. 5.21, wherein a random mapping of challenges is used to design the permutation block, wherein each challenge is divided into k sections, each contains L bits, those sections are then permuted. We consider three cases, here, $k = 4$, 8 and 16.

(b) **Substitution-Based Challenge Obfuscation**

In this case, each challenge is divided into two sections, each contains one byte (8 bits), these bytes are then substituted using the AES substitution functions, and applied to the PUF.

(c) **Substitution-Based Response Obfuscation**

Each challenge is used as seed for a 16-bit LFSR which generates 16 corresponding challenges. The obtained challenges are applied to the PUF sequentially to generate a 16-bit response in each case. Each response is then divided into two sections, each contains one byte (8 bits), these bytes are then substituted using the AES substitution functions.

In each of the above cases (A, B and C), 18000 CRPs are generated, both SVM and ANN techniques are then used to develop a software model of the design. The machine learning results are shown in Figs. 5.22 and 5.23.

The results in Fig. 5.22 show that the permutation can help reduce the prediction accuracy of SVM and ANN machine learning models by up to 25 and 10%, respectively, to achieve the maximum reduction, the number of bits in each permuted block should be 1 (i.e. k = 16). The results in Fig. 5.23 show that substitution is a more effective approach, compared to permutation, especially when applied on the response bits.

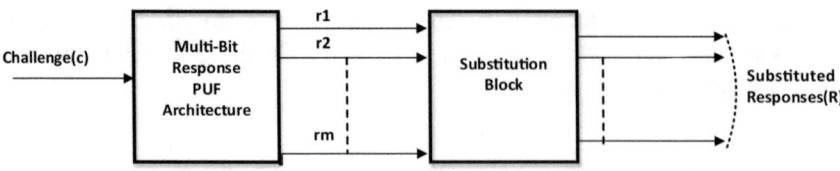

Fig. 5.21 Architecture of substituted response PUFs

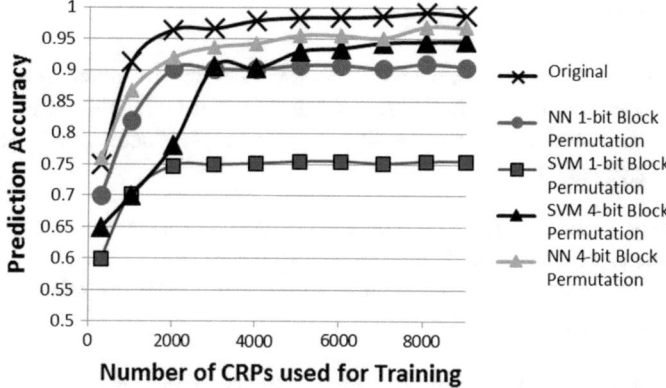

Fig. 5.22 The impact of permutation on the unpredictability of arbiter PUFs

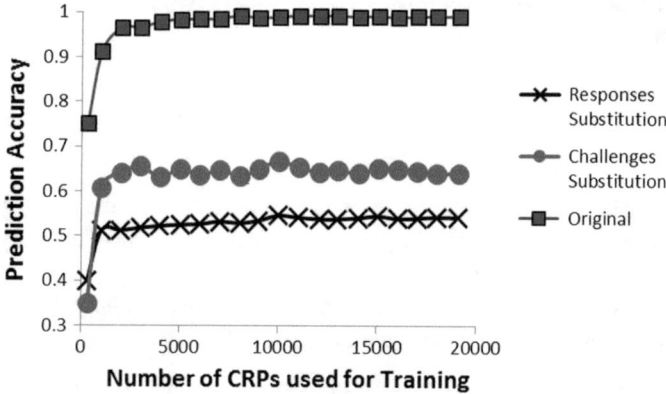

Fig. 5.23 The impact of substitution on the unpredictability of arbiter PUFs

5.10   Side Channel Attacks

Side channel attacks exploit information leaked from the physical implementation of a cryptographic primitive, such as power consumption, timing delays and electromagnetic noise. These attacks can be classified based on their implementation strategies into two main categories, the first requires information on the internal structure and operation of the hardware implementation of the system under attack such as fault injection methods, and the second type may not necessarily need such information, for example differential power analysis.

In this section, we will explain the principles of the main side channel attacks on PUFs.

5.10.1   Fault Injection Attacks

An example of such attack has been described in [29], where authors introduced the concept of 'Repeatability', which refers to the short-term reliability of a PUF that is affected by temporal noise sources (e.g. power supply fluctuations). This attack consists of two stages.

(a)   An experimental stage, which consists of injecting noise into the PUF circuit and measuring the repeatability of its responses.
(b)   Statistical analysis stage which infers the internal structure of the device and produces a mathematical clone.

We are going to briefly explain how this attack works using the arbiter PUF depicted in Fig. 5.12.

Let γ be the minimum resolution time of the arbiter (i.e. the minimum difference in the arrival times of the two input signals that can be detected by the arbiter).

The difference of the arrival times Δt is the sum of two independent components as follows:

$$\Delta t = \Delta t_V + \Delta t_N \tag{5.18}$$

where

Δt_V is caused by the variation-induced delay difference of the two signal paths associated with each challenge, this component is dependent only on manufacturing variations.

Δt_N is caused by random temporal noise such as power supply fluctuation and changes in the ambient temperature.

Both of these components can be considered to have Gaussian distributions according to the central limit theorem because they are generated by complex processes (i.e. random variations or random noise).

The core principle of this attack is to systematically inject noise and then measure its impact on the response bits. To achieve this, a Gaussian noise component is injected into the power supply, and the response of the arbiter PUF is measured repeatedly for the same challenge.

The experiment is repeated for different mean values of the noise signal, in each case, the repeatability metric (REP) is estimated as the fraction of the responses which evaluate to '1' for a specific challenge, therefore $REP \in [0, 1]$.

REP can be computed by integrating the probability distribution function of Δt as follows:

$$REP = \frac{1}{2} erfc \left(\frac{\gamma - \Delta t}{\sqrt{2}\sigma_N} \right) \tag{5.19}$$

where

Δt is the mean value of the time difference
σ_N is the standard deviation of the noise signal

This can be rewritten as

$$REP = \frac{1}{2}erfc\left(\frac{\delta - \Delta t_V - \Delta t_N}{\sqrt{2}\sigma_N}\right) \qquad (5.20)$$

This means

$$\Delta t_V = \gamma + \Delta t_N - \sqrt{2}\sigma_N erfc^{-1}(2REP) \qquad (5.21)$$

Equation (5.21) crucially indicates that repeatability measures can be used to compute Δt_V for each challenge, hence gaining insight into the internal structure of the arbiter circuits.

The relationship between REP and Δt_V is linear when $REP \in [0.1, 0.9]$. We have already established from Sect. 5.7.6 that the relationship between Δt and the delay of the internal stages of the arbiter $(\delta_1^{1/0}, \delta_2^{1/0}, \ldots \delta_k^{1/0})$ is also a linear function, this means the whole PUF degrades to a linear system. To produce a model for the a k-stage arbiter PUF (as the one shown in Fig. 5.13), a set of $(k + 1)$ CRPs and their corresponding repeatability figures are evaluated, then the latter are used to solve a system of linear equation, whose unknowns are $(\delta_1^{1/0}, \delta_2^{1/0}, \ldots \delta_k^{1/0})$.

This attack may not be sufficient on its own to produce a mathematical clone of the PUF in the case of RO PUF, but it can help reduce the complexity of machine learning based attacks as shown in [29], in addition, its effectiveness can be further enhanced by increasing the ratio of unstable CRPs by inducing more significant changes in the environment conditions [1].

5.10.2 Power Analysis Attacks

There are a number of examples in the literature wherein power consumption information are exploited to wage more effective mathematical cloning attacks [30, 31].

In this section, we are going to explain the generic procedure of these attacks using the arbiter PUF (Fig. 5.12). The arbiter circuit consumes more power when generating a logic '1' response than when generating a logic '0', therefore by applying power tracing to this circuit, one can predict the output of the PUF.

The presence of noise makes it difficult to obtain specific response bits from the power measurement, therefore, machine learning algorithms that require a set of challenge response pairs such as ANN and SVM cannot be employed in this case. Other techniques, which can be applicable in this context, include evolution strategies techniques, which start with a random model of the PUF, then iteratively refine this model using available data (e.g. power consumption information) until an

---

Step 1: Initialize Arbiter *the arbiter is set to its initial state (which is assumed to be the same for all challenges)*

Step 2: Experimental Data Collection

  For i=1, K   *number of challenges to be used*

    Start

           Apply challenge (c_i)

           $DB_i = [c_i, p_i]$ * record power consumed for at the end of each evaluation

           (p_i))   along with its corresponding challenge in a data base DB

           value*

    End

Step 3: Model Building using Evolution Strategies Technique

    1. Generate a random model of the PUF called the parent model **(PM)**.
    2. Compute the prediction quality of the parent model using the measurement data base **(DB)**
    3. If the prediction quality of the parent model is high go to (7) otherwise go to (4)
    4. Generate a number of children models **(CM)** by randomly modifying the delay values of the parent model **PM**
    5. Evaluate the prediction quality of all children models and choose a new parent model which give the most accurate prediction
    6. Go to (2)
    7. Exit

Fig. 5.24 Power analysis attack using an evolution strategies learning algorithm

accurate model is derived. A generic procedure of this attack that is based on the use of ES learning algorithm is shown in Fig. 5.24.

The authors of [31] have shown that it is possible to build a model of a 128-bit arbiter PUF, implemented using 45 nm CMOS technology, with a prediction accuracy more than 90% using 15000 challenges and their associated power traces, their experiment also assumed a noise equivalent to that generated by one hundred switching registers in the same technology node.

It should be noted here that the number of challenges needed for this attack is dependent on the noise level, therefore, if the latter can be reduced, then it will be possible to significantly reduce the number of power measurements required [31].

There are other forms of power analysis attacks which use Hamming distance as a correlation metric, these approaches exploit the fact that the power consumed by a memory element (e.g. a shift register) when its content changes is dependent on the Hamming distance between the new content and the old one. Let us give an example of this. Consider a 8-bit register that contains the binary string (00001000), if we are to write a new word (11111000), we would need to draw current from the power supply to change the four most significant bits from '0' to '1', on the other hand, if we are to write (11001000), we would draw less current as we only need to make two transitions. This principle has been successfully exploited to deduce information form secure-sketch-based PUF designs [32].

5.10.3   Helper Data Leakage

Cryptographic key generation is one of the primary applications of PUF technology, in this case only one challenge is applied to the PUF to generate a response which can be used to derive an encryption key. However, environmental variations and deep submicron noise decrease the reliability of the PUF responses, which may lead to generating the wrong key.

Helper data algorithms are used to compensate for these effects in order to meet the stringent stability and entropy requirements of encryption keys.

However, helper data can be manipulated to deduce sensitive information [14], we will explain the principle of this type of threats using a key recovery attack on RO PUFs [4].

The attack exploits helper data used in the index masking scheme(see Sect. 4.7.1 for more details) the latter is a reliability enhancement technique proposed in [33], in which the designer can preselect a subset of ring oscillators to be compared such that the difference in the respective frequencies of each selected pairs is larger than any possible frequency shift due to noise or environment variations. The helper data in this case consist of the indexes of the pairs of ring oscillators, which can be compared. This pairing information is generated during enrollment phase at post-fabrication stage, and subsequently stored in a on-chip memory. The attack described here assumes the adversary is able to manipulate the helper data and observe the output of the cryptographic block, furthermore, the attack assumes that erroneous re-construction of the encryption key leads to an observable increase in failure rate at the output of the cryptographic block. The essence of this attack is to swap the memory positions of the helper data vectors corresponding to two response bits, then measure the output of the cryptographic block. If this change leads to a sudden increase in the failure rate, it means the two affected response bits are not the same. If no change in the output can be observed, it means that the two response bits are identical. By repeating this process for each pair of response bits, the adversary will be able to construct a set of linear equations which relate each pair of the PUF response bits. Solving these equations will give two possible solutions, one of them is the key, which can be easily identified as it will lead to a smaller error rate at the output of the cryptographic block.

The general procedure of this attack is described in Fig. 5.25.

To explain the last step in the above algorithm, let us consider the example below:

Example 5.8 A helper data manipulation attack has been carried out on a key generation scheme, which uses an RO PUF with a 4-bit response. The results of the experimental phase of the attack are summarised in Table 5.8. Deduce the two possible encryption keys, assuming the used key is the same as the response of the PUF.

Solution:

Table 5.8 can be used to generate 12 linear equations which indicate the relationship between different response bits, in this case we will only need four independent equation to find a solution as follows

---

Step 1 Parameters Definition

      i, j: integer

      R: the total number of response bits

      P_{FAIL}: failure rate at the output of the cryptographic block

      r_i: response bits position i

      r_j : response bit position j

      d_i, d_j: Helper data bits which indicate which pair of ROs corresponds to r_i, r_j respectively

      H_{ij}: hypothesis which correspond to the swapping of the helper data indices

Step 2 Experimental Data Collection

      For i=1:R , j:R

      Start

            Swap di, dj

            Generate a key

            Evaluate P_{FAIL}

            If P_{FAIL} is unchanged than Hij =1 (i.e. $r_j = r_j$)

            If P_{FAIL} is increased than Hij =0 (i.e. $r_j \neq r_j$)

            Store Hij

      End

Step 3 Data Analysis

Analyse Hij to deduce the key

When the all response bits are matched, only two possible values remain for the secret key, only one of these can minimise the P_{FAIL}

Fig. 5.25 Helper data analysis attack

Table 5.8 Experimental results of a helper data manipulation attack on RO PUF

H_{12}	H_{13}	H_{14}	H_{21}	H_{23}	H_{24}	H_{31}	H_{32}	H_{34}	H_{41}	H_{42}	H_{43}
1	0	0	1	0	0	0	0	1	1	0	1

$$r_1 = r_2 \quad because \; (H_{12} = H_{21} = 1)$$

$$r_3 = r_4 \quad because \; (H_{34} = H_{43} = 1)$$

$$r_1 \neq r_3 \quad because \; (H_{13} = H_{31} = 0)$$

$$r_2 \neq r_4 \quad becasue \; (H_{24} = H_{43} = 0)$$

From the above equations, one can deduce two possible solutions $r_1r_2r_3r_4 = 0011 \; or \; 1100$, one of them is the key.

5.11 Countermeasures of Side Channel Attack

Techniques to enhance the security of cryptographic systems against side channel attacks have been extensively researched. Countermeasures for power analysis attacks aim to conceal the correlation between power traces and sensitive data, this is done by noise insertion [34], random clock frequency [35], randomisation of the instruction streams [36] or random insertion of dummy instructions into the execution sequence of algorithms [37]. There are other types of techniques which aim to stop or reduce the leakage of side channel information in the first place; examples of these approaches include the use of asynchronous design to reduce electromagnetic (EM) radiation. This technique has been shown to be effective in improving the resilience of RO PUF against EM-analysis [3], wherein the authors showed that using asynchronous counters to design the RO PUF makes it harder to wage a successful EM-attacks, the same work also included a list of circuit-based countermeasures specifically for RO PUFs.

Power balanced logic is another technique that aims to remove the correlation between the data being processed on chip and its corresponding power consumption, one method to achieve this is to use (m-of n) circuits [38, 39]. For fault injection attacks, noise insertion can be an effective defence mechanism [34].

Although the effectiveness of the above-mentioned methods in protecting classic cryptographic cores against side channel attacks has been rigorously investigated, there is still a great deal of work to be done to evaluate their applicability in the case of PUFs, this is still an active area of research.

5.12 Physical Cloning Attacks

So far, we have discussed attacks which aim to create a mathematical model of the PUF, such attacks may not be very useful in highly secure applications such as smart cards and e-passports, wherein a clone can only be practically used if it has the same form factor (i.e. physical size and shape) as the original device. In such cases, creating a physical clone of the device may be the only practical threat; such a clone should have the same functionalities, capabilities and physical properties of the original device.

Physical cloning aims to clone physically unclonable functions, although this may seem contradictory, the advancement of reverse engineering technique has made physical cloning a possibility [3, 13].

This type of attacks generally consists of two stages: characterisation and emulation. The aim of the first stage is to deduce the challenge/ response behaviour of a PUF circuit from its electrical and physical characteristics (delay, radiation, output consistency, etc.). The goal of the second stage is to construct a physical implementation, which replicates the learned behaviour. We will now discuss existing techniques used in each stage.

5.12.1 Physical Characterisation Techniques of PUF

There are a number of methods with which the integrated circuits of a PUF can be characterised. These include side channel analysis such as electromagnetic radiation and power consumption. For example, the authors of [3] de-capsulated an RO PUF implemented on an FPGA board and held on-die EM measurements. These are then used to deduce the frequency ranges of each ring oscillator in the design, this allowed full characterisation of the design.

Photonic emission (PE) in CMOS is another technique that was used to physi-cally characterise an arbiter PUF [2]. PE phenomenon occurs during the switching of CMOS devices. For example, let us consider the NMOS transistor in Fig. (5.26). In this case, electrons travelling through the conducting channel are accelerated by the electric field between the source and the drain, when they arrive at the pinch-off region at the drain edge of the channel, their kinetic energy can be released in the

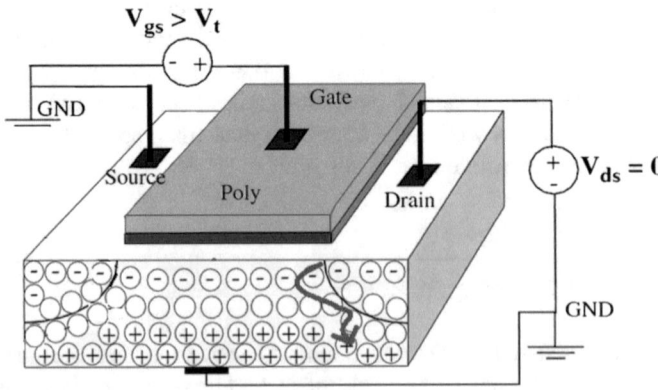

Fig. 5.26 Photonic emission in NMOS transistors

form of photons (illustrated as a red arrow in Fig. 5.26). The wavelength of the generated light particles is dependent on the energy of the accelerated carries, it can range from the visible to the near-infrared wavelengths in some CMOS technologies. This phenomenon is also present in PMOS transistors, but the generated photons have less energy, due the lower mobility of holes compared to electrons, which mean the emitted photons are much harder to observe.

The authors of [2] measure the time difference between enabling the arbiter PUF and the photon emission at the last stage (detected from the back side of the chip), this allowed them to estimate the delay of each individual path in the design. The experiment was repeated for different challenges; they showed it is possible to estimate the delay of all the stages of a 128-bit arbiter PUF with 1161 challenges, which is significantly less than the number of CRPs required to carry out a mathematical cloning attack.

Photonic emission analysis has also been proven effective in capturing the start-up values of an SRAM PUF [13]. Data remanence decay in volatile memories is another mechanism, which has been proven effective in characterising the SRAM PUF [40]. This method consists of two phases, in the first stage, an attacker overwrites the start-up values of the SRAM memory with known contents, and then he gradually reduces the supply voltage and captures the memory content iteratively. In the second stage, he analyses the results to deduce the PUF responses, in principles, the SRAM cells that remain stable have a start-up value equivalent to the stored known content, while others which change their status have a start-up value equal to the reverse of the forcibly stored contents.

5.12.2 Physical Emulation Methods of PUFs

The previous section showed that physical characterisation of PUF circuits is possible, however creating a physical clone is a much harder problem especially at nanoscale CMOS technologies wherein the limitations imposed by quantum mechanics make it impossible to accurately control the circuit fabrication process, a step necessary for creating an exact replica of an integrated circuit.

Despite this, the authors of [13] managed to produce a physical clone of an SRAM PUF, they used a technique called *focused ion beam circuit edit* (*FIB CE*) to modify the layout of a volatile memory in order to give it the same start-up value of an SRAM PUF.

This technique was applied to the memory block in Atmel ATmega328P device, the latter was first de-packaged and thinned using an automated mechanical polishing machine.

The FIB CE process includes milling the back side of the device to the n-well, this is followed by another thinning stage at the precise locations of the circuit to be modified (see Fig. 5.27). In order to give an SRAM memory cell a specific start-up value, its transistors are trimmed individually to modify their dynamic behaviour or can be even removed completely so the cell is incapable of storing a particular logic value.

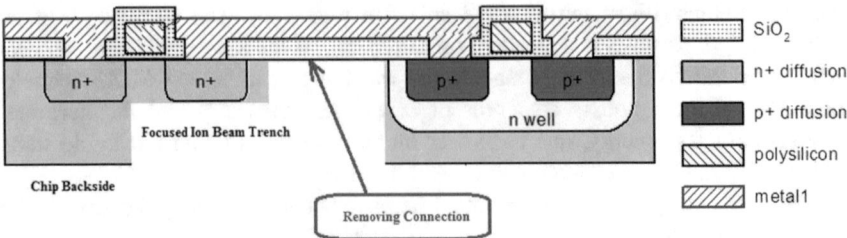

Fig. 5.27 Focused ion beam circuit edit

5.13   Countermeasures of Physical Cloning Attacks

The advancement of reverse engineering techniques, such as PEA and FIB, means that it is feasible for an adversary to acquire detailed information on the layout of a silicon chip and potentially deduce its intended functionality. This means, he may be able to reproduce a physical replica of the original device if he has access to the same or closely related implementation technology. This makes it hard to devise countermeasures that can achieve 100% protection against physical emulation (i.e. the second stage of a physical attack).

Therefore a more effective approach, in this case, is to develop solutions to prevent physical characterisation, i.e. the first stage of the attack, such methods may include the countermeasures for side channel analysis described in Sect. 5.11 and also industry standard techniques such as memory scrambling [41, 42].

5.14   Comparative Analysis of Attacks and Countermeasures

Table 5.9 shows a summary of the attacks discussed in this chapter, which includes the adversary type and the compromised security metric in each case.

Mathematical cloning attacks are the most likely threat as they can be carried out by all three types of adversaries. They mainly compromise the *unpredictability* property of a PUF by constructing a model, which can predict responses for all future challenges; they can also affect its *Uniqueness*, if such a model is used to construct another device with the same challenge/response behaviour.

These attacks have been successfully carried out on most existing PUFs with large challenge/response space such as delay-based designs. Other architecture such as memory-based PUFs are inherently resilient to mathematical modelling as the total number of their CRPs is typically less than that required by most machine learning algorithms.

Table 5.9 Summary of security attacks on physically unclonable functions

Attacks type		Adversary type			Attacked PUF designs	Compromised security metrics			
		Snooping	Black Box	White Box		physical unclonability	Unpredictability	Randomness	Uniqueness
Mathematical Cloning		X	X	X	Arbiter, TCO, and RO [43, 44]		X		X
Side Channel Attacks	Faults Injection,		X	X	Arbiter [29] RO [45] SRAM [40]		X		X
	Power analysis		X	X			X		X
	Electromagnetic Radiation analysis		X	X	RO [3]		X		X
	Helper Data Leakage			X	RO PUF [4]		X		X
Physical Cloning				X	Arbiter [2] SRAM [13]	X	X	X	X

In terms of countermeasures for these attacks, there are numerous solutions which increase the complexity of behavioural modelling such as (e.g. permutation) or makes it unfeasible (e.g. MTR and controlled PUFs), there are currently no reported techniques to prevent a mathematical cloning by a white box adversary for most existing silicon-based PUF architectures, therefore new design solutions are needed.

Side channel attacks require a more capable adversary, especially those based on fault injection and helper data leakage, but they have the same effects on the security metrics as mathematical cloning.

Physical cloning attacks can only be carried out by a white box adversary, their effects on the security of the PUF are the most severe as they compromise all four properties, including randomness.

Although attacks, which aim to reduce the randomness of a PUF device, have not been reported in the literature, it is not unimaginable for a white box attacker to modify the physical structure of a PUF circuit to undermine its randomness in order to weaken its unpredictability and render it more amenable to cloning.

There is ongoing research to devise effective countermeasures for reported side channel and physical attacks, we have discussed some of the potential solutions, but a great deal of work is still needed in this area.

An interesting observation from the summary table is that most reported attacks in the literature target SRAM or delay-based PUFs, therefore, a proper evaluation of the robustness of other emerging and existing PUF designs against these attacks is required.

5.15　Conclusions and Learned Lessons

A summary of important lessons is provided below:

- There are four metrics which need to be used collectively to assess the security of a PUF design, namely *randomness, unpredictability, uniqueness and physical Unclonability*.
- There are three types of adversaries which can be considered when formally reasoning about the security of PUFs, namely: a snooping, a black box and a white box adversary. The first can only listen to communication from/to a PUF circuit, the second has a physical access to the chip incorporating the PUF and can carry out non-invasive attacks and the third can carry out both semi and fully invasive procedures such as micro-probing and reverse engineering.
- There are a number of attacks on physically unclonable functions, which pose a serious risk to their potential use as a new root of trust in security protocols. These attacks can generally be classified into three categories according to the means by which the attack is carried out, mathematical modelling using machine learning algorithms, side channel analysis and physical cloning.

- Successful implementations of these attacks have been reported on the majority of popular designs such as memory and delay-based designs, however, the feasibility of some of these attacks on newly emerging designs is yet to be established.
- There are a number of existing and emerging countermeasures which render some of these attacks more difficult or impossible in some cases.
- The countermeasures for modelling attacks include obfuscating the challenges and/or responses, and increasing the time needed to collect sufficient data for machine learning algorithms, however these countermeasures are not useful in the case of a white box adversary, and this leaves the following problem unresolved: *How to protect a PUF against machine learning attacks waged by a white box adversary?* Potential solutions are likely to consist of new PUF designs which are inherently resilient to modelling, whether or not this is feasible remains to be seen.
- There is still a great deal of work that needs to be done in order to develop and evaluate effective countermeasures against side channel and physical attacks.
- The question '*how secure a PUF implementation is*' can only be answered with respect to a specific threat model, which can be formulated based on the intended application and the most likely type of adversaries.

5.16   Problems

5.1. Table 5.10 shows the challenge/response behaviour of two PUF designs, each of which has 3-bit challenges and 6-bit responses.

  (a) Calculate the uniformity and output entropy for each design
  (b) Which of these two PUFs, if any, can be used as true random bit generator?

Table 5.10 Challenge/response behaviour for two PUF designs

PUF challenge	Design 1	Design 2
000	010110	011000
001	011100	000100
010	110100	110001
011	000101	100001
100	010011	110010
101	110101	111010
110	000101	010000
111	000101	011101

5.2. Table 5.11 shows the challenge/response behaviour of a PUF that has 3-bit challenges and 4-bit responses.

(a) Use the Hamming distance metric $HDT(t, e)$ to evaluate the unpredictability of this PUF using $t = 2$.

(b) A permutation approach is implemented on Design 1 to enhance its unpredictability; it consists of a left circular shift of each challenge before applying it to the PUF. Evaluate the unpredictability of the resulting design using $t = 2$.

5.3. The minimum readout time (MTR) approach is a preventative countermeasure, which aims to severely restrict the ability of an adversary to collect sufficient number of challenge/response pairs needed to build a mathematical clone. Assuming a black box adversary needs at least 10 thousand CRPs for this attack:

(a) How long will it take him to carry out this attack assuming he needs one microsecond to obtain one challenge/response pair?

(b) One method to implement MTR is by inserting a logic delay at the output of the PUF, and estimate the approximate value of this delay in order to increase the time needed for this attack to 30 days.

(c) Assume that the above delay block needs to be implemented using a 65 nm technology, in which the delay of an inverter is 1.1 ns, estimate how many of these inverters will be needed.

5.4. A helper data leakage attack has been carried out on a key generation scheme based on a 4-bit RO PUF, the results of the experimental phase of the attack are summarised in Table 5.12. Analyse the data and deduce the possible keys, assuming the used key is the same as the response of the PUF.

Table 5.11 Challenge/response behaviour

Challenge	Responses
000	0110
001	0001
010	1100
011	1010
100	1001
101	1111
110	1000
111	1110

Table 5.12 Experimental results of a helper data manipulation attack on RO PUF

H_{12}	H_{13}	H_{14}	H_{21}	H_{23}	H_{24}	H_{31}	H_{32}	H_{34}	H_{41}	H_{42}	H_{43}
0	0	1	0	1	0	0	1	0	1	0	0

Table 5.13 A summary of the metrics of three different PUFs

Design	$\left\{\mu(HDT(t,e)), \frac{\sigma(HDT(t,e))}{\mu(HDT(t,e))}\right\}$	Uniqueness (%)	Reliability (%)	Entropy
PUF1	(0.47, 0.3)	45	97	0.46
PUF2	(0.49, 0.01)	54	90	0.48
PUF3	(0.48, 0.1)	48	94	0.5

5.5. The AES S-box is used to enhance security of a PUF design against mathematical cloning. It is assumed that the PUF has a reliability metric of 98.875%, which is estimated based on the expected number of errors due environmental variations and other sources of noise. It is also assumed that the PUF has 8-bit challenges and 8-bit responses

(a) Estimate the reliability of the resulting enhanced PUF design if the S-box is placed on its input.
(b) Estimate the reliability of the resulting enhanced PUF design if the S-box is placed on its output.
(c) Based on your answers to the above, explain whether or not the position of the S-box has any effect on the reliability of the resulting design.

5.6. Table 5.13 includes a summary of the security metrics of three PUF designs. Study these data then choose a suitable PUF design for each of the following applications:

(a) Cryptographic Keys Generation.
(b) True Random Number Generator.
(c) Mutual Authentication Protocols.

References

1. J. Delvaux, I. Verbauwhede, Fault injection modeling attacks on 65 nm Arbiter and RO sum PUFs via environmental changes. IEEE Trans. Circ. Syst. I Regul. Pap. **61**, 1701–1713 (2014)
2. S. Tajik, E. Dietz, S. Frohmann, J.-P. Seifert, D. Nedospasov, C. Helfmeier, et al., Physical Characterization of Arbiter PUFs, in ed. by L. Batina, M. Robshaw. *Cryptographic Hardware and Embedded Systems—CHES 2014: 16th International Workshop, Busan, South Korea, September 23–26, 2014, Proceedings* (Springer, Berlin, 2014), pp. 493–509

3. D. Merli, D. Schuster, F. Stumpf, G. Sigl, Semi-invasive EM attack on FPGA RO PUFs and countermeasures. Presented at the proceedings of the workshop on embedded systems security, Taipei, Taiwan, 2011
4. J. Delvaux, I. Verbauwhede, Key-recovery attacks on various RO PUF constructions via helper data manipulation, in *2014 Design, Automation & Test in Europe Conference & Exhibition (DATE)* (2014), pp. 1–6
5. E.B.B. Morris, J. Dworkin, J.R. Nechvatal, J. Foti, L.E. Bassham, E.Roback, J.F. Dray Jr., *Advanced Encryption Standard (AES), Federal Inf. Process. Stds. (NIST FIPS)—197* (2001, July, 2017). Available: https://www.nist.gov/publications/advanced-encryption-standard-aes
6. L. Daihyun, J.W. Lee, B. Gassend, G.E. Suh, M.V. Dijk, S. Devadas, Extracting secret keys from integrated circuits. IEEE Trans. Very Large Scale Integr. VLSI Syst. **13**, 1200–1205 (2005)
7. V.G.A. Maiti, P. Schaumont, A systematic method to evaluate and compare the performance of physical unclonable functions. IACR ePrint **657**, 245–267 (2013)
8. A. Rukhin, J. Soto, J. Nechvatal, M. Smid, E. Barker, S. Leigh, M. Levenson,M. Vangel, D. Banks, A. Heckert, J. Dray, S. Vo, A statistical test suite for random and pseudorandom number generators for cryptographic applications. Special Publication 800-22 Revision 1a, NIST, Apr 2010
9. G. Marsaglia, *The Marsaglia random number CDROM including the diehard battery of tests of randomness.* Available: http://www.stat.fsu.edu/pub/diehard/
10. C.E. Shannon, A mathematical theory of communication. Bell Syst. Tech. J. **27**, 623–656 (1948)
11. C.E. Shannon, A mathematical theory of communication. Bell Syst. Tech. J. **27**, 379–423 (1948)
12. A. Renyi, On measures of entropy and information, in *Proceedings of the Fourth Berkeley Symposium on Mathematical Statistics and Probability, Volume 1: Contributions to the Theory of Statistics, Berkeley, CA* (1961), pp. 547–561
13. C. Helfmeier, C. Boit, D. Nedospasov, J.P. Seifert, Cloning physically unclonable functions, in *2013 IEEE International Symposium on Hardware-Oriented Security and Trust (HOST)* (2013), pp. 1–6
14. J. Delvaux, D. Gu, D. Schellekens, I. Verbauwhede, Helper data algorithms for PUF-based key generation: overview and analysis. IEEE Trans. Comput. Aided Des. Integr. Circ. Syst. **34**, 889–902 (2015)
15. S. Katzenbeisser, Ü. Kocabaş, V. Rožić, A.-R. Sadeghi, I. Verbauwhede, C. Wachsmann, PUFs: myth, fact or busted? A security evaluation of physically unclonable functions (PUFs) cast in silicon, in ed. by E. Prouff, P. Schaumont, *Cryptographic Hardware and Embedded Systems—CHES 2012: 14th International Workshop, Leuven, Belgium, September 9–12, 2012. Proceedings* (Berlin, Heidelberg, 2012), pp. 283–301
16. U. Rührmair, J. Sölter, PUF modeling attacks: an introduction and overview, in *2014 Design, Automation & Test in Europe Conference & Exhibition (DATE)* (2014), pp. 1–6
17. P.H. Nguyen, D.P. Sahoo, R.S. Chakraborty, D. Mukhopadhyay, Security analysis of Arbiter PUF and its lightweight compositions under predictability test. ACM Trans. Des. Autom. Electron. Syst. **22**, 1–28 (2016)
18. A. Arbit, Y. Oren, A. Wool, Toward practical public key anti-counterfeiting for low-cost EPC tags, in *2011 IEEE International Conference on RFID* (2011), pp. 184–191
19. I. Steinwart, A. Christmann, *Support vector machines* (Springer, New York, 2008)
20. M.S. Mispan, B. Halak, M. Zwolinski, Lightweight obfuscation techniques for modeling attacks resistant PUFs. Presented at the 2nd international verification and security workshop: IVSW 2017. IEEE, 2017
21. S.O. Haykin, *Neural Networks and Learning Machines* (Pearson Education, 2011)
22. E.D. Karnin, A simple procedure for pruning back-propagation trained neural networks. IEEE Trans. Neural Netw. **1**, 239–242 (1990)

23. M.S. Mispan, B. Halak, Z. Chen, M. Zwolinski, TCO-PUF: a subthreshold physical unclonable function, in 2015 *11th Conference on Ph.D. Research in Microelectronics and Electronics (PRIME)* (2015), pp. 105–108

24. M. Majzoobi, M. Rostami, F. Koushanfar, D.S. Wallach, S. Devadas, Slender PUF protocol: a lightweight, robust, and secure authentication by substring matching, in *2012 IEEE Symposium on Security and Privacy Workshops* (2012), pp. 33–44

25. Y. Gao, G. Li, H. Ma, S.F. Al-Sarawi, O. Kavehei, D. Abbott, et al. Obfuscated challenge-response: a secure lightweight authentication mechanism for PUF-based pervasive devices, in *2016 IEEE International Conference on Pervasive Computing and Communication Workshops (PerCom Workshops)* (2016), pp. 1–6

26. R. Plaga, F. Koob, A formal definition and a new security mechanism of physical unclonable functions. Presented at the Proceedings of the 16th international GI/ITG conference on Measurement, Modelling, and Evaluation of Computing Systems and Dependability and Fault Tolerance, Kaiserslautern, Germany, 2012

27. B. Gassend, D. Clarke, M.V. Dijk, S. Devadas, Controlled physical random functions, in *2002 Proceedings on 18th Annual Computer Security Applications Conference* (2002), pp. 149–160

28. W. Trappe, R. Howard, R.S. Moore, Low-energy security: limits and opportunities in the internet of things. IEEE Secur. Priv. **13**, 14–21 (2015)

29. J. Delvaux, I. Verbauwhede, Side channel modeling attacks on 65 nm arbiter PUFs exploiting CMOS device noise, in *2013 IEEE International Symposium on Hardware-Oriented Security and Trust (HOST)* (2013), pp. 137–142

30. U.R. Ahmed Mahmoud, M. Majzoobi, F. Koushanfar, Combined modeling and side channel attacks on strong PUFs. IACR Cryptol. ePrint Arch. **632** (2013)

31. G.T. Becker, R. Kumar, Active and passive side-channel attacks on delay based PUF designs. Cryptoeprint (2014)

32. D. Merli, D. Schuster, F. Stumpf, G. Sigl, Side-channel analysis of PUFs and fuzzy extractors, in ed. by J.M. McCune, B. Balacheff, A. Perrig, A.-R. Sadeghi, A. Sasse, Y. Beres, *Proceedings on Trust and Trustworthy Computing: 4th International Conference, TRUST 2011, Pittsburgh, PA, USA, June 22–24, 2011* (Springer, Berlin, 2011), pp. 33-47

33. M.D. Yu, S. Devadas, Secure and robust error correction for physical unclonable functions. IEEE Des. Test Comput. **27**, 48–65 (2010)

34. J. Jaffe, P. Kocher, B. Jun Differential power analysis. *CHES* (1999)

35. M.L. Akkar, Power analysis, what is now possible. *ASIACRYPT* (2000)

36. P. Grabher, J. Großschädl, D. Page, Non-deterministic processors: FPGA-based analysis of area, performance and security, in *Proceedings of the 4th Workshop on Embedded Systems Security, Grenoble, France* (2009)

37. C. Clavier, J.S. Coron, N. Dabbous, Differential power analysis in the presence of hardware countermeasures, in *Proceedings of the Second International Workshop on Cryptographic Hardware and Embedded Systems,* vol. 1965 (LNCS, 2000), pp. 252–263

38. B. Halak, J. Murphy, A. Yakovlev, Power balanced circuits for leakage-power-attacks resilient design, in *Science and Information Conference (SAI)* (2015), pp. 1178–1183

39. NewcastleUniversity, Cryptographic processing and processors, U.K. Patent Appl. No. 0719455.8, 4 Oct 2007

40. S. Zeitouni, Y. Oren, C. Wachsmann, P. Koeberl, A.R. Sadeghi, Remanence decay side-channel: the PUF case. IEEE Trans. Inf. Forensics Secur. **11**, 1106–1116 (2016)

41. M.I. Neagu, L. Miclea, S. Manich, Improving security in cache memory by power efficient scrambling technique. IET Comput. Digital Tech. **9**, 283–292 (2015)

42. G.T. Becker, The gap between promise and reality: on the insecurity of XOR Arbiter PUFs, in ed. by T. Güneysu, H. Handschuh, *Proceedings on Cryptographic Hardware and Embedded Systems—CHES 2015: 17th International Workshop, Saint-Malo, France, September 13–16, 2015* (Springer, Berlin, 2015), pp. 535–555

43. J. Daemen, V. Rijmen, *The Design of Rijndael: AES—The Advanced Encryption Standard* (Springer, Berlin, 2013)

44. M. Backes, A. Kate, A. Patra, Computational verifiable secret sharing revisited, in ed. by D.H. Lee, X. Wang, *Proceedings on Advances in Cryptology—ASIACRYPT 2011: 17th International Conference on the Theory and Application of Cryptology and Information Security, Seoul, South Korea, December 4–8, 2011* (Springer, Berlin, 2011), pp. 590–609

45. C. Alexander, G. Roy, A. Asenov, Random-Dopant-induced drain current variation in nano-MOSFETs: a three-dimensional self-consistent Monte Carlo simulation study using (Ab initio)Ionized impurity scattering. IEEE Trans. Electron Devices **55**, 3251–3258 (2008)

Hardware-Based Security Applications of Physically Unclonable Functions

6

6.1 Introduction

The applications of physically unclonable functions are versatile ranging from secure cryptographic key storage to advance security protocols such as oblivious transfer schemes.

This chapter aims to:

(1) Explain how PUF technology can be used to securely generate and store cryptographic keys.
(2) Discuss the principles of PUF-based entity authentication schemes.
(3) Explain how PUF technology can be employed to construct hardware-assisted security protocols.
(4) Outline the principles of PUF-based secure sensors design.
(5) Explain how PUFs can be used to develop anti-counterfeit solutions and anti-tamper integrated circuits.

It hoped that this chapter will give the reader an in-depth understanding of the existing PUF applications, their design requirements and outstanding challenges.

6.2 Chapter Overview

The remainder of this chapter is organised as follows. Section 6.3 illustrates, with a detailed case study, how to design and implement a PUF-based key generation scheme. Section 6.4 explains the principles of existing PUF-based authentication protocols and discusses the motivation behind their usage in this type of applications. Section 6.5 outlines the principles of hardware-assisted security protocols and explains with numerical examples how PUFs can be used in this context. Section 6.6 examines the use of this technology to construct secure sensors and gives

© Springer International Publishing AG, part of Springer Nature 2018
B. Halak, *Physically Unclonable Functions*,
https://doi.org/10.1007/978-3-319-76804-5_6

detailed design guidelines. Section 6.7 outlines the principles of PUF-based anti-counterfeiting techniques. Section 6.8 briefly explains how PUF technology can be used to design anti-tamper integrated circuits. The chapter concludes with a summary of learned lessons in Sect. 6.9. Finally, problems and exercises are provided in Sect. 6.10.

6.3   Cryptographic Key Generation

6.3.1   Motivation

Security is one of the fundamental requirements of electronic systems that deal with sensitive information; especially, these used in smart cards, mobile phones and secure data centres. Therefore, these systems should be capable of protecting data confidentiality, verifying information integrity and running other necessary security-related functions. Such requirements are typically achieved using cryptographic primitives such as encryption cores and hash functions. These cryptographic blocks normally require a secret key that should be only known to authorised persons or devices.

Typically, secret keys are generated on the device from a root key, which must be unpredictable (i.e. cannot be easily guessed) and securely stored (i.e. cannot be accessed by an adversary).

Nowadays root keys are typically generated outside the chip during the manufacturing stage of silicon devices, and subsequently stored in an on-chip memory. This process is usually referred to as key provisioning, once completed; the on-chip cryptographic block can provide security services (e.g. data encryption, device authentication and IP protection) to the applications running on the device.

There is typically an application key provisioning function, which allows the operating system and other software to derive their own application keys based on the root key.

There are generally two types of storage medium used for the secret keys. The first is one-time programmable (OTP) memories, wherein a fuse or an anti-fuse locks each stored bit; in this case, the data need to be written during chip manufacturing and cannot be changed later on. Another approach for storing the key is the use of non-volatile memories such as Flash, FRAM and NRAM.

The above described key provision approach suffers from two problems:

(a) First, the use of OTP memories for root key storage is handled by the device manufacturers, which adds extra costs to the fabrication process, it can also pose a security risk as the vast majority of devices are fabricated in third-party facilities, which cannot always be trusted.

(b) Second, storing the keys in a non-volatile memory makes it vulnerable to read-outs attacks by malicious software, an example of such attacks is described in [1], which demonstrates how a malware can gain an unauthorised access to a device's memory. Another security risk is data remanence attacks [2, 3], which makes it feasible for secret information to be deduced from memories even if they have been erased.

PUF technology provides an alternative approach for root key generation; it consists of using the response of a PUF to construct a root key. This approach removes the need for storing the root key in a non-volatile memory, as it can be generated only when it is needed; this makes it more resilient to memory read-out and data remanence attacks. PUF-based key generation also improves the security of the key provisioning process because the device manufacturer will no longer be required to handle key generation/injection. Examples of PUF key generation schemes in the literature include [4–6].

6.3.2 PUF Design Requirements

This section examines more closely the requirements a PUF design should satisfy in order to be used as the basis of a key generation scheme, these can be summarised as follows:

(a) High Reliability: which means the PUF response remains the same regardless of noise and environment variations, this ensures stable key generation, without which decryption would be very difficult, if not impossible. It is typically the case that PUF responses are not perfectly reproducible; therefore there may be a need to use additional circuity for errors correction.
(b) Uniqueness: the key should be unique for each electronic system, so that if one device is compromised, the others remain secure; ideally, the uniqueness should be 50%.
(c) Randomness: the key should be random so it is impossible for an adversary to guess it or compute it, it is typically the case that PUF responses are not uniformly distributed, and therefore, there may be a need for an additional circuitry in order to compress enough entropy in a PUF-generated key; ideally, a random response should have a uniformity of 50%.

6.3.3 PUF-Based Key Generation Process

The process of generating a cryptographic key from a PUF consists of two main stages shown in Fig. 6.1:

(a) Setup Stage:

This is carried out only once by the developer/seller of a device, it includes the following procedure:

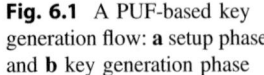

Fig. 6.1 A PUF-based key generation flow: **a** setup phase and **b** key generation phase

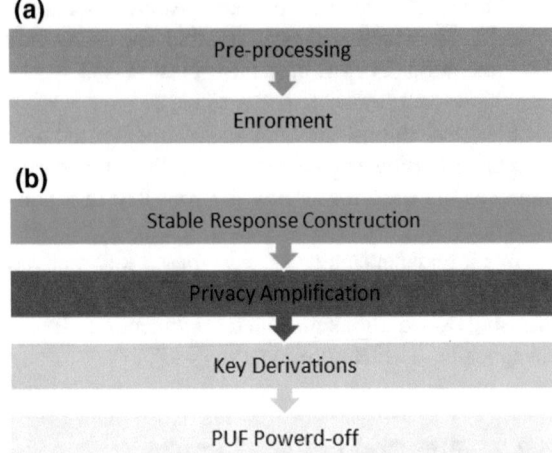

(a)

Pre-processing

Enrorment

(b)

Stable Response Construction

Privacy Amplification

Key Derivations

PUF Powerd-off

(1) Pre-processing: the aim of this stage is to estimate the maximum bit error rate of the PUF response when the same challenge is applied under different environment and noise conditions. It can also include the use of a number of approaches to reduce this error rate such as ageing acceleration, and reliable bit selection (see Chap. 4 for more details on this). This process takes place at the design stage; it requires fabrication of test chips and evaluation of the reliability of the PUF design under realistic operating conditions.

(2) Helper Data Generation: wherein helper data are generated for each PUF implementation using one of the fuzzy extractor schemes explained in Chap. 4, and these data are then stored into each PUF-carrying chip, in principles, helper data should leak no information on the key, so it can be stored on chip without any additional protection measures. This step takes place after the chip fabrication process is completed.

(3) Device Enrolment: the root key stored on each device is read-out and stored securely by an authentication authority to be used for secure communication with the device in question at a later stage.

(b) Key Generation:
This is carried out in the field on demand (i.e. whenever a key is required), it consists of the following steps:

(1) Stable Response Construction wherein the PUF is provided with a specific challenge or simply powered-on (e.g. SRAM PUF), its output is then fed into an error correction block to generate a reliable response as shown in Chap. 4.

(2) Privacy Amplification wherein the generated response is applied to an entropy compression block (e.g. a hash function) to enhance its randomness.

(3) Key Derivations: The output of the last step is used to derive one or multiple keys for different security tasks (encryption, identification, etc.).

(4) Finally, the PUF is powered-off, so that its output is no longer accessible.

This technique has a number of advantages over the existing approaches; first, it ensures that the root key is only available temporarily when it is needed (e.g. during an encryption/decryption operation). This makes it harder for an adversary to exploit side channel information to deduce the key. Second, there will be no need for the chip manufacturer to handle key provisioning, which reduces fabrication costs and improved security. Third, the use of PUF technology makes it easier to obtain a unique key for each device, which makes the key provisioning process significantly easier and less prone to errors.

6.3.4 Design Case Study: PUF-Based Key Generation

This section illustrates the design process of a PUF-based 256-bit key generation scheme. A memory-based PUF design is adopted due to the stability of their response in the presence of environmental variations [7]. A 2 MB embedded SRAM on an Altera DE2-115 FPGA board is used in this experiment [8], it has a 20-bit address input line and a 16-bit data output, and these are used as challenge/response pairs. To simplify the design process, only one source of environment noise is considered, namely, power supply variations. The design process is explained in detail below:

(a) *Setup Stage*

1. Reliability Characterisation:

   The purpose of this step is to estimate the worst-case bit error rate in a given PUF response, in order to identify a suitable error correction scheme. Ideally, a comprehensive assessment should be carried out that includes all possible sources of noise as explained in Chap. 3. In this example, for simplicity, the power supply fluctuation is the only source of noise considered. The experimental procedure is outlined below:

   (1) The SRAM memory is powered-up.
   (2) The start-up values of the first 100 addressable locations are captured.
   (3) The data is then transmitted from the FPGA to a PC using the RS232 serial port.
   (4) For each addressable location, the experiment is repeated at different supply voltages ranging between 95% to 105% of the nominal value.
   (5) The maximum bit error rate for each 16-bit response is computed by calculating its Hamming distance from the response at the nominal voltage supply and finding the maximum value in each case.

   The results indicated that the responses can be categorised into two types according to their maximum bit error rates, 64 responses had a maximum bit error rate of 5×10^{-4}. The remaining responses had a higher error rate $> 10^{-3}$.

2. Reliable Bit Selection:
   This step aims to identify the response bits that are least likely to flip due to voltage fluctuations, in order to reduce the cost of the required error correction scheme. To do this, the PUF challenges (i.e. the SRAM memory addresses) that give a maximum of that give a maximum of error rate of 5 x 10^{-4} are chosen.

3. Identifying the number of raw PUF response:
   The purpose of this step is to determine the minimum number of raw response bits (r) needed to generate a k-bit cryptographic key from a PUF device that has an an estimated error rate (P_e). In this example, code-offset scheme is used for key recovery (see Chap. 4 for details on this) wherein an error correcting block (n, m, t) is used (where n is the length of the code words (typically $r = n$), m is the length of the datawords and t is the maximum number of errors that can be corrected). $\frac{m}{n}$ is referred to as the code rate CR.
   For the purpose of stable key generation, one needs to ensure that the decoding error of the chosen error correction code does not exceed 10^{-6}, this generates an m-bit stable data vector. The latter is applied to a privacy amplifier which generates a k-bit key. The minimal amount of compression that needs to be applied by the privacy amplifier is referred to as the secrecy rate SR [9].
   The above discussion allows us to deduce the following relationship the minimum number of required PUF response bits (r) and the the length of the key that need to be generated (k)

$$r = \frac{k}{CR \times SR} \tag{6.1}$$

In this case study, a Hamming code (15, 11, 1) is chosen, which meets the requirement of the decoding error < 10^{-6}. To demonstrate this, we recall Eq. (4.4) from Chap. 4.
In this example, we have:

$$P_e = 5 \times 10^{-4}, n = 15, t = 1, \text{therefore}$$

$$P_{ECC} = 1 - ((1 - P_e)^{15} + 15 \times P_e \times ((1 - P_e)^{14}))$$

By substituting $P_e = 5 \times 10^{-4}$, we get $P_{ECC} = 2.6 \times 10^{-7}$
This confirms that the chosen Hamming codes has an acceptable decoding error, so it can be adopted for the PUF key generation scheme. The coding rate of this code can be computed as follows:

$$CR = \frac{m}{n} = \frac{11}{15}$$

To compute the maximum achievable secrecy rate, the ContextTree Weighting method is used [10]. To do this, the mutual information between

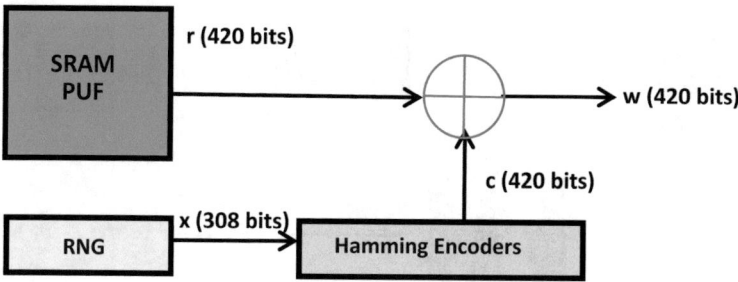

Fig. 6.2 Helper data generation using code-offset secure sketch

multiple readings of the selected memory space are estimated, the results indicated an average SR of 0.85.

The minimum number of raw PUF response needed to generate a 256-bit cryptographic key can now be computed using Eq. (6.1) as follows:

$$r = \frac{k}{CR \times SR} = \frac{256}{\frac{11}{15} \times 0.85} \approx 410 \; bits$$

In this case, the PUF responses will be generated from the start-up values of the memory addresses identified in step a.2, each of which has 16-bits value, and only the first 15 will be used from each response (to allow the use of the Hamming code chosen). This means, the total number of addresses that need to be read (i.e. the number of challenges that need to be applied to the PUF) is given as follows: $\frac{410}{15} = 27.3$. This effectively means the contents of 28 addresses need to be read-out, which gives 420 bits.

4. Helper Data Generation:

In this step, a helper data vector is generated using a code-offset scheme described in Chap. 4 [11], the detailed process is explained as follows (see Fig. 6.2):

(1) A 420-bit raw response bits are generated by reading out the start-up value of a 28-memory addresses.
(2) A random binary string x whose length is 308-bits is generated using a random number generator (RNG).
(3) x is divided into 28 data vectors, each is 11-bit long. These are applied in parallel to Hamming encoder to generate 28 valid codewords, these are concatenated to generate (c) that is a 420-bit long vector.
(4) c is XORed with the PUF response (r) to generate a helper data vector w.

(b) **Key Generation**

This stage consists of two steps, a response reconstruction that is done using the recovery stage of the code-offset secure sketch scheme shown in Fig. 6.3. The second step is privacy amplification which aims to increase the entropy of the generated key. The details are provided below:

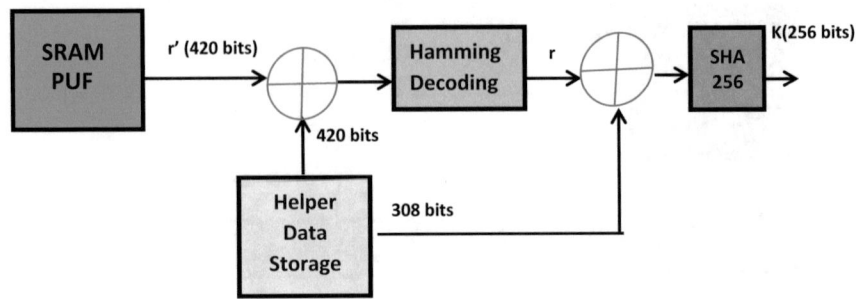

Fig. 6.3 Key generation using code-offset secure sketch

(a) A 420-bit response vector is generated by reading out the start-up values of the same 28 memory locations used in the previous stage.

(b) Each 15-bit response is XORed with its corresponding helper data w to obtain noisy code vector $c' = r' \oplus w$.

(c) The 28 obtained noisy codewords are applied to Hamming decoders to generate 28 * 11-bit data vectors.

(d) The outputs from all decoders are concatenated, then XORed with the corresponding helper data to generate a 308-bit stable response.

(e) Finally, the stable response bits are inputted into SHA-256 Hash function block to generate a 256 key.

The above scheme was implemented on an Altera DE2-115 FPGA board, it was tested by carrying out repeated readings of PUF responses and estimating their respective Hamming distances from the original response. The design has successfully generated the original response in each case in the presence of supply voltage variations. The obtained key has a uniformity of 49.6% which is very close to the ideal value of 50%. It worth noting here, that in practical applications, more rigorous analysis are needed to identify the realistic bit error rate at the output of the PUF, according to the operating conditions and expected sources of noise. In extreme scenarios, if the error rate is large, the cost of error correction may be prohibitive. In these cases, one solution is to adopt a multi-level optimization approach to improve the reliability of the PUF. This may include using robust circuit design techniques, preprocessing approaches (e.g. aging acceleration), in addition to error correction circuitry.

6.4 Entity Authentication

6.4.1 Motivation

Nowadays, there are numerous applications that require physical objects to be identified and/or authenticated. Examples of use cases of entity authentication include tracking goods in manufacturing applications, secure products transfer in the consumer services industry, gaining access to subscribed TV channels, cash withdrawal in banking systems and border control using e-passports.

In each of the above examples, the identity of a physical object needs to be established before a service can be offered. For example, in the case of cash withdrawal, the bank needs to ensure that the information stored on the card is authentic and consistent with those in its database before it can release the money. In this context, we refer to the authentication authority as the '*verifier*', for example, the government is the verifier of e-passports and the bank is the verifier of the credit cards. Also, the entity or the physical object to be verified is sometimes referred to as the '*prover*'.

When a physical entity wants to authenticate itself to a verifier, it needs to provide the following:

(a) Corroborative evidence of its claimed identity which could have only been generated by the entity itself.
(b) A proof that entity was actively involved in generating this evidence at the time of the authentication.

To achieve this, the entity needs to convince the authenticating authority that it has exclusive access to secret information (e.g. a binary code), and demonstrate it can generate a proof of this secret as and when requested.

Generally, there are two stages of an entity authentication scheme:

(a) Identity Provisioning: wherein each device obtains a unique identity such as a serial number or a binary code.
(b) Verification Phase: wherein the verifier validates the identity of each entity.

Existing entity authentication approaches are based on the use of encryption algorithms or keyed hash functions. We will give an example of these based on the ISO/IEC 9798-2 standard, which uses the symmetric challenge-response technique, and it works as follows:

At the provisioning stage,

(1) Verifier (V) or a trusted third party gives each entity (E), a unique secret key (k) and a unique identifier (ID).
(2) Verifier stores this information in its database.

At the verification stage,

(1) Entity starts the process by sending its (ID) to the verifier.
(2) Verifier looks up the (k) corresponding to a specific E with the unique (ID).
(3) Verifier sends a random number (nonce) to the entity.
(4) Entity encrypts (nonce) using the shared key (k), generates a response (ne) and sends it back to the verifier.
(5) Verifier decrypts the received response (ne) using the shared key (k).
(6) If the decryption process generates the same (nonce), then the entity can be authenticated because it has proved its possession of the key (k); otherwise, the authentication process fails.

The disadvantages of the above approach are as follows:

(1) It requires assigning each entity a unique binary key that needs to satisfy strict randomness conditions (it should not be easy for an adversary to guess it). This process, referred to as key provisioning, should take place before devices are deployed to the field or given to customers; this poses technical and logistical challenges, and leads to an increased production cost. For example, the vast majority of silicon chips are manufactured by third parties who cannot necessarily be trusted for key provisioning or may not have the technical skills to generate a unique random key for each device.

(2) It requires the implementation of an encryption algorithm or a keyed hash function on each entity, this may be unaffordable in resources-constrained systems such as low-end IoT devices and RFID [12, 13].

PUF technology provides an alternative approach to designing entity authentication schemes, which can overcome the above disadvantages, as it allows each device to generate its own unique key, hence remove the need for key assignments. In addition, it makes it feasible to construct authentication protocols that do not require the implementation of resources-demanding encryption algorithms, as will be shown in the following sections.

6.4.2   PUF Design Requirements

The core principle of PUF-based authentication schemes is to use the unique challenge/response behaviour of each PUF instance to generate an inherent identifier for each physical entity. Therefore, there are a number of qualities a PUF design should possess in order to be suitable for this type of applications, as summarised below:

(a) Mathematical unclonability which prevents an eavesdropping adversary from building a software clone of the device if he manages to collect enough challenge/response pairs.

(b) High Reliability: which means the PUF can generate the same response for a specific challenge regardless of noise and environment variations, failure to do this can lead to a denial of service.

(c) Uniqueness: the challenge/response behaviour of the PUF needs to be unique for each chip to allow differentiating multiple devices in a network.

It is worth noting that the above qualities of the PUF can only be evaluated with respect to the requirements of a specific authentication protocol. In other words, some protocols limit the number of challenge/response pairs used, which means it will not be possible for a snooping adversary to construct a model of the PUF using mashie learning algorithms, this in turns relaxes the requirement of mathematical unclonability. Other applications can tolerate a higher rate of authentication failure, which means reliability requirement can be relaxed.

6.4.3   Basic PUF Unilateral Authentication Scheme

This protocol can be used for a unilateral authentication, wherein a central authority acts as the verifier and PUF-embedded devices as the entities. There are a number of variations of this protocol that have been proposed, but they all have the following basic procedure shown in Fig. 6.4 [14]:

(a) *Enrolment*

   (1) The verifier or a trusted third party embeds in each entity a PUF circuit and gives it a unique identifier (ID).
   (2) Verifier applies a large number of challenges on each PUF instance and records the corresponding responses.
   (3) Verifier creates a secure database, in which he stores the IDs for all entities with their corresponding PUF challenge/response pairs.

(b) *Verification*

   (1) Entity starts the process by sending its ID to the verifier.
   (2) Verifier looks up the challenge/response vectors which correspond to the received (ID).
   (3) Verifier sends a challenge (C) to entity.
   (4) Entity applies the received challenge (C) to its PUF and sends back the generated response (R').
   (5) Verifier compares the received response (R') with that stored in its database (R), if they are equal then the entity is authenticated; otherwise, the authentication request is denied.
   (6) Verifier deletes the challenge/response pair used in the above process to prevent replay attacks.

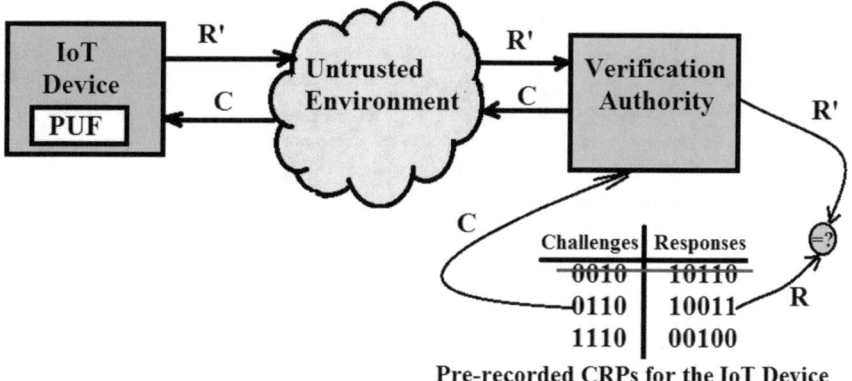

Fig. 6.4 Basic PUF-based authentication process

6.4.4   The Principles of Fuzzy Authentication

As discussed previously, PUF responses are not perfectly reproducible when challenged multiple times, this is due to temporal noise. This makes it hard to construct a PUF-based authentication scheme that requires 100% recovery of originally recorded responses. Instead, a '*Fuzzy*' authentication approach is used, in which the verifying authority accepts responses as long as their Hamming distance from the original response is less than a predefine threshold.

However, if such threshold is large, it may lead to false identification, this is because the range of allowed responses from two different physical entities can overlap, which makes it harder to rely on the PUF to provide a unique identifier for each chip. This can be best illustrated using the intra- and inter-Hamming distance distributions of a typical PUF as shown in Fig. 6.5.

Although the variations in PUF responses, caused by temporal noise (intra-Hamming distance), are smaller than those between responses from different chips (inter-Hamming distance), there is still an overlap between these two distributions. This effectively means, it is possible for temporal noise to turn one PUF response of an entity into a valid response of another entity, which leads to false identification.

In order to choose the appropriate identification threshold, one needs to evaluate the following two metrics:

(1) False Acceptance Rate (*FAR*): which refers to the probability that a PUF of a physical entity generating a response identical to that of a PUF implemented on another entity. This leads to false identification.
(2) False Rejection rate (*FRR*) is the probability of a genuine physical entity generating an invalid response due to temporal noise.

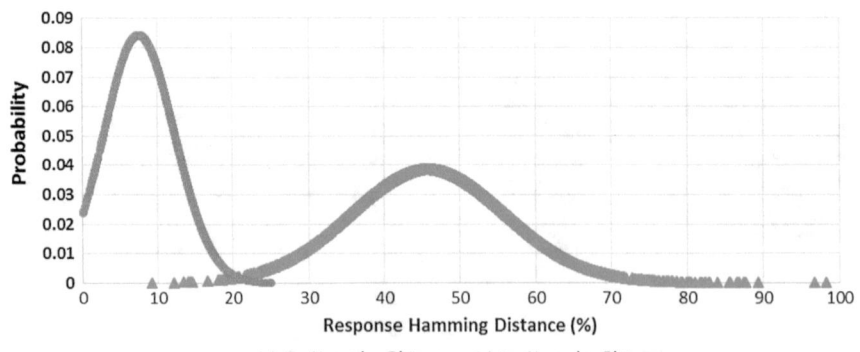

Fig. 6.5 Intra- and inter-hamming distance distributions of a 32-bit arbiter PUF

FRR and *FAR* can be estimated using the HD_{Inter} and HD_{Intra} as follows [15]:

$$FRR = 1 - \sum_{i=0}^{t} \binom{M}{i} \left(\hat{p}_{HD\_Intra}\right)^{i} \left(1 - \hat{p}_{HD\_Intra}\right)^{M-i} \qquad (6.2)$$

$$FAR = \sum_{i=0}^{t} \binom{M}{i} \left(\hat{p}_{HD\_Inter}\right)^{i} \left(1 - \hat{p}_{HD\_Inter}\right)^{M-i} \qquad (6.3)$$

wherein

M is the number of PUF output bits (i.e. the bit length of its response per challenge).

t the identification threshold or the maximum number of bit error in the PUF response that can be tolerated, i.e. do not cause a false rejection.

$\hat{p}_{IntraPQ}, \hat{p}_{IntrePQ}$ are the binomial probability estimators of HD_{Intra} and HD_{Inter}, respectively.

It is worth noting here that the above equations are only applicable if HD_{Intra} and HD_{Inter} have binomial distributions.

Ideally, both *FRR* and *FAR* need to be minimised; however, Eqs. (6.2) and (6.3) indicate that in order to minimise *FRR*, *t* needs to be as large as possible so the bit flips caused by temporal noise do not lead to rejecting a genuine response. On the other hand, increasing *t* can aggravate *FAR* as it increases the probability of false identification. Therefore, in practice, *t* is chosen to balance *FRR* and *FAR*. In other words, to make them equal. This value of *t* is referred to as the equal error threshold (t_{EER}). For discreet distributions, it may not be possible to find such a value, instead t_{EER} is computed as follows:

$$t_{EER} = \text{argmin}_{t}\{\max\{FAR(t), FRR(t)\}\} \qquad (6.4)$$

And the equal error rate, in this case, will be

$$EER = \max\{FAR(t), FRR(t)\} \qquad (6.5)$$

To illustrate the impact the choice of a threshold value has on the admission and rejection rates, the following numerical examples are given below:

Example 6.1 The challenge/response behaviour of a PUF is given in Table 6.1. This design is used in the authentication protocol depicted in Fig. 6.4. The PUF is expected to operate under four different environmental conditions (i.e. power supply fluctuations and ambient temperature). The nominal conditions at which the

Table 6.1 PUF challenge/response behaviour under different environment conditions

Environment conditions	Challenges			
	$c_0 = 00$	$c_1 = 01$	$c_2 = 10$	$c_3 = 11$
$T = 25\,°C, V_{dd} = 1v$	00000000	00000111	00111000	11111111
$T = 75\,°C, V_{dd} = 1v$	00000000	00000111	00110000	11111000
$T = 25\,°C, V_{dd} = 1.2v$	00000001	10000111	00100000	01111000
$T = 75\,°C, V_{dd} = 1.2v$	00000001	11000111	00000000	00111000

PUF was enrolled at are $(T = 25\ °C, V_{dd} = 1v)$. It is assumed that all challenges listed in Table 6.1 have equal probabilities, and all the listed environment conditions are also equally probable.

(1) Compute the false rejection and admission rates assuming the verifier cannot tolerate any errors, i.e. the threshold t = 0.
(2) Repeat the same computations above but with a threshold t = 1.
(3) Which of the threshold values produce the minimum equal error rate (EER)?

   Solution:

(1) The PUF has four valid responses listed next the nominal conditions $(T = 25\ °C, V_{dd} = 1v)$. To compute the false rejection rate, one needs to look for cases wherein the Hamming distance between the PUF output and any of the four valid responses is more than 0. There are eight of these cases written in a bold font in Table 6.2. As it is assumed all table entries have the same probability, the false rejection rate can be computed as follows:

$$FRR = \frac{8}{16} = 0.5$$

To compute the false admission rate, one needs to look for cases wherein the PUF output has deviated from its expected response and generated another valid response. There are two of such cases written in an italic underlined font in Table 6.2, therefore

Table 6.2 False rejection (bold fonts) and false admission (italic font) cases (t = 0)

Environment conditions	Challenges			
	$c_0 = 00$	$c_1 = 01$	$c_2 = 10$	$c_3 = 11$
$T = 25\,°C, V_{dd} = 1v$	00000000	00000111	00111000	11111111
$T = 75\,°C, V_{dd} = 1v$	00000000	00000111	**00110000**	**11111000**
$T = 25\,°C, V_{dd} = 1.2v$	**00000001**	**10000111**	**00100000**	**01111000**
$T = 75\,°C, V_{dd} = 1.2v$	**00000001**	**11000111**	*00000000*	*00111000*

Table 6.3 False rejection (bold font) and false admission (italic underlined font) cases (t = 1)

Environment conditions	Challenges			
	$c_0 = 00$	$c_1 = 01$	$c_2 = 10$	$c_3 = 11$
$T = 25\,°C, V_{dd} = 1v$	00000000	00000111	00111000	11111111
$T = 75\,°C, V_{dd} = 1v$	00000000	00000111	00110000	**11111000**
$T = 25\,°C, V_{dd} = 1.2v$	00000001	10000111	*00100000*	*01111000*
$T = 75\,°C, V_{dd} = 1.2v$	00000001	**11000111**	*00000000*	*00111000*

$$FAR = \frac{2}{16} = 0.125$$

(2) In the case of a threshold value (t = 1), the verifier can tolerate up to 1-bit error per response; therefore, to compute the false rejection rate, one needs to look for cases wherein the Hamming distance between the PUF output and any of the four valid responses is more than 1. There are two such cases in this example as written in a bold font in Table 6.3, therefore

$$FRR = \frac{2}{16} = 0.125$$

To compute the false admission rate, one needs to look for cases wherein the PUF output has deviated from its expected response by more than 1 bit, but it has a Hamming distance from another valid response that is less than 2. There are four of such cases written in an italic underlined font in Table 6.3, therefore

$$FAR = \frac{4}{16} = 0.25$$

(3) The equal error rate is estimated as $EER(t) = \max\{FAR(t), FRR(t)\}$
So

$$EER(t = 0) = 0.5$$

$$EER(t = 1) = 0.25$$

Therefore, t = 1 gives the minimum equal error rate in this case.

Example 6.2 A PUF design is chosen to be used in the authentication protocol depicted in Fig. 6.4. The false rejection/admission rates are estimated for different threshold values, and the results are listed in Table 6.4. Analyse the given data and answer the following questions:

Table 6.4 Characterization results of false rejection/admission rates of a PUF design

Authentication threshold (t)	False rejection rate (FRR)	False admission rate (FAR)
4	4.2×10^{-4}	1.2×10^{-5}
6	3.6×10^{-5}	2.2×10^{-5}
8	8.6×10^{-6}	1.8×10^{-4}

(1) What is the best threshold value that minimises the probability of a denial of service?
(2) What is the best threshold value that minimises the probability of a forged PUF being authenticated?
(3) What is the best threshold value that reduces both of the above risks?

Solution:

(1) Denial of service can be caused by rejecting to authenticate a genuine device; therefore to decrease this probability, one needs to minimise the false rejection rate, and therefore, in this case, t = 8 is the best threshold value to achieve this.
(2) The probability of authenticating a forged PUF is related to the false admission rate, this can be minimised by choosing t = 4.
(3) t = 6 can reduce the probabilities of both risks.

6.4.5 Advance PUF Authentication Schemes

The basic PUF authentication described above does not require the use of encryption, which significantly reduces implementation costs of the prover hardware compared to classic schemes; however, it has two main drawbacks, namely:

(1) It has a major logistical disadvantage, as each device needs to be individually enrolled; this is not a scalable process and may not be practical for some applications with a large number of devices such as wireless sensor networks.
(2) It is vulnerable to machine modelling attacks, wherein an eavesdropping adversary collects sufficient challenge/response pairs to construct a mathematical model of the PUF. This may be prevented by enhancing the PUF design with obfuscation circuitry (as seen in Chap. 5), but this comes at extra implementation costs.

Slender PUF was proposed in [16] to overcome some of the disadvantages of the basic PUF authentication protocol, in particular, the need for a verifier to store large numbers of challenge/response pairs for each device. It works as follows:

(a) Enrolment Phase:

   (1) Verifier or a trusted third party embeds in each entity a PUF circuit and gives it a unique identifier (ID).

   (2) Verifier applies a large number of challenges on each PUF instance and records the corresponding responses, and then it constructs a mathematical model of for each PUF using machine learning algorithms.

   (3) Verifier stores the IDs for all entities with their corresponding PUF software models.

(b) Verification Phase:

The verification stage is depicted in Fig. 6.6; it consists of the following steps

   (1) Entity starts the process by sending its ID to the verifier and a random binary vector (nonce e).

   (2) Verifier checks the ID and sends back another random binary vector (nonce v). Then, both entity and verifier concatenate the two nonce vectors to generate (e, v).

   (3) Entity uses a previously agreed upon pseudorandom function (G) to generate a challenge based on the seed, and it then applies the challenge c to its PUF instance and generates a response r of length m, where $r = \{b_0, b_1, \ldots b_m\}$.

   (4) Verifier uses the same random function to generate the same challenge based on the seed, and then applies it to the PUF model corresponding to the entity's identifier (ID); this generates a response $r' = \{b'_0, b'_1, \ldots b'_m\}$.

   (5) Entity selects a substring of the response of length s where $sub(r) = \{b_i, b_j, \ldots b_s\}$ and sends it back to the verifier along with the indexes of the chosen bits, i.e. $\{i, j, \ldots, s\}$.

   (6) Verifier computes the Hamming distance between the received substring $sub(r)$ and the corresponding substring from its generated response $sub(r')$. If the Hamming distance is smaller than the authentication threshold (t), then the entity is authenticated; otherwise, the authentication request is denied.

The Slender protocol makes it difficult to wage modelling attacks by a snooping adversary as the challenges are not transmitted in the clear and only partial responses are exchanged; however, it still requires the characterization of each PUF instance in order to build a model.

Another advanced authentication protocol has been implemented in [17], wherein a partial challenge is sent by the verifier to a prover device embedding the PUF, and subsequently padded with a random pattern generated by a random number generator to make up a full-length challenge before being applied to the PUF. The server (i.e. the verifier) uses a challenge recovery mechanism to generate an emulated response to compare with the received response.

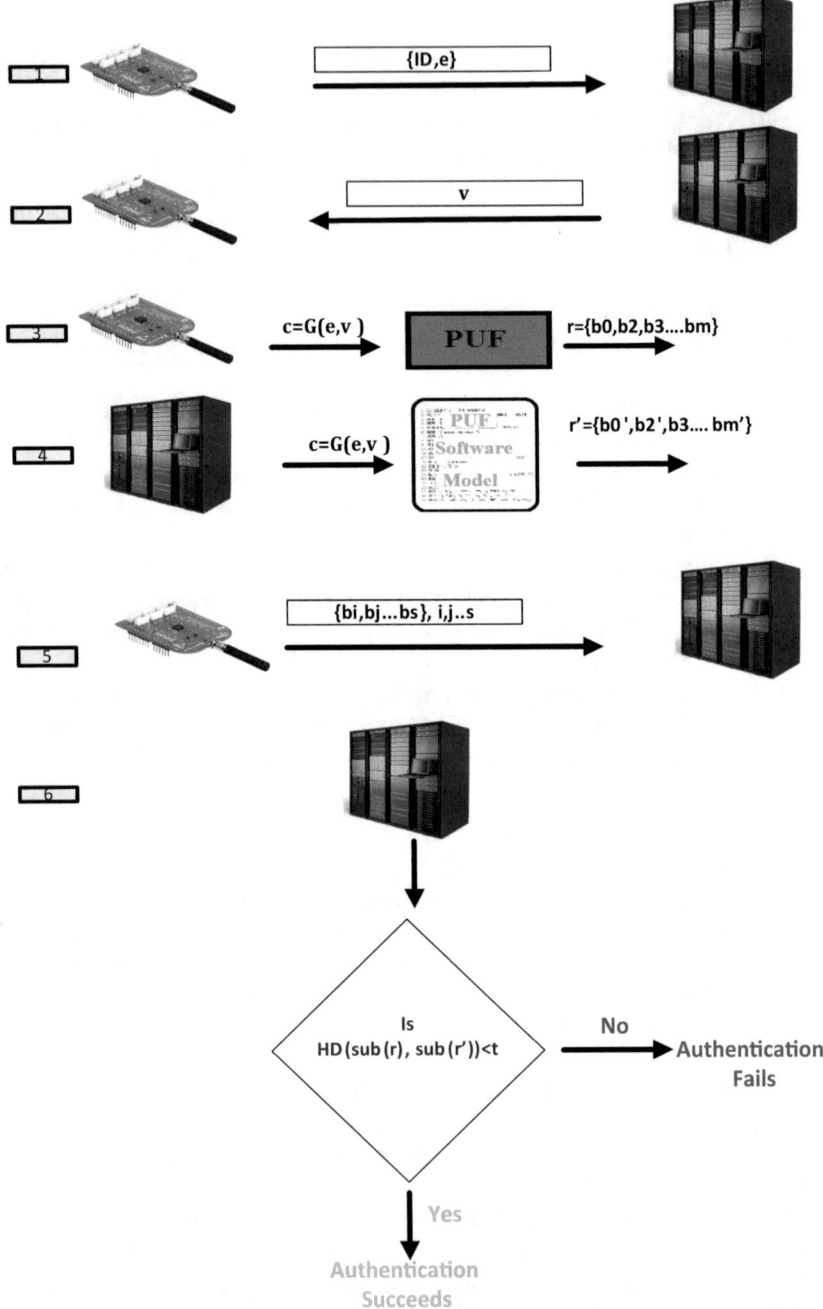

Fig. 6.6 Slender PUF authentication protocol

Both of the above approaches might increase the authentication time as well as consume significant computing resources on the verifier side. One might argue, however, this is not a problem since the verifier has always been assumed rich in resources.

A slightly different approach is discussed in [18], wherein the challenges are divided into two subsets: valid and invalid; the former are called the secret-challenges or (the s-challenges), and the number of valid challenges in this set should not be sufficient to allow an adversary to build a mathematical model.

6.5 Hardware-Assisted Cryptographic Protocols

6.5.1 Motivation

In a secure multiparty computation, a number of parties need to carry out joint communication based on their private individual inputs; there are a great many applications which require the use of such type of computation, including private data mining, electronic voting and anonymous transactions. The security requirements in such cases are threefold:

(1) No single party can learn about the inputs of other parties using the output of the protocol.
(2) The input of each party should be independent of those of other parties.
(3) The output is only accessible by authorised parties as described by the protocol.

However, protocols that meet such requirements do not typically have efficient implementations in practice [19]. This has given rise to hardware-assisted cryptographic protocols; the latter rely on the use of tamper-proof hardware tokens to help achieve the strong security guarantees set in the Canetti's universal composition (UC) framework [20]. In this type of protocols, the trust between communicating parties is established through the exchange of trusted hardware tokens. One of the first examples of such schemes was presented in [21], wherein government-issued signature cards are used to generate public/private key pairs for digital signature schemes. A second example includes the use of smart cards-based schemes in private data mining applications [19]. A third example consists of the use of secure memory device that restricts the number of times the memory contents can be accessed, this functionality is particularly useful in applications such as intellectual properties protection of software [22].

In this context, physically unclonable functions have been proposed as a suitable technology for the construction of hardware-assisted secure protocols due to their complex challenge/response behaviours, which make them intrinsically more tamper-resistant than other hardware tokens that rely on digitally stored information (e.g. smart cards or secure memory devices) [23].

This section describes three Hardware-assisted PUF-based cryptographic protocols, namely: key exchange, oblivious transfer and bit commitment.

6.5.2 PUF-Based Key Exchange (KE) Schemes

These protocols are used to allow the sharing of secret keys among two or more of authorised parties for subsequent use in secure communication sessions. One of the earliest PUF-based KE protocols was proposed in [24], it assumes the existence of a secure physical transfer mechanism of the PUF. A similar approach was presented in [25]; it is based on the use of universal composition framework [20], wherein a key-carrying PUF is physically transferred from a sender to a receiver; in this case, a one-sided authentication is required in order to prevent an adversary who gain a temporary access to the PUF from impersonating the sender. A more elaborate scheme was presented in [26], which will be explained in more details below. It is also illustrated in Fig. 6.7.

(1) Bob applies two challenges c1, c2 to a PUF and obtains corresponding responses r1, r2.
(2) The PUF is then physically transferred to Alice.
(3) Alice acknowledges the reception of the PUF using an insecure binary channel.
(4) Bob sends (c1, r1) and c2 to Alice.
(5) Alice applies c1 to the received PUF and get r1'.
(6) Alice compares r1' and r1. If they are different, she terminates the communication; otherwise, she proceeds to step 7.
(7) Alice applies the challenge c2 to generate r2.
(8) Bob and Alice use r2 to derive a shared secret.

6.5.3 Oblivious Transfer (OT)

An Oblivious Transfer protocol in its simplest form enables a sender to transfer one or multiple data items to a receiver while remaining oblivious to what pieces of information have been sent(if any). The OT protocol has been first proposed in [27] wherein a sender transfers a message to a receiver with 50% probability without knowing whether or not his message is delivered. Another form of this scheme, called 1-of-2 oblivious transfer, was proposed later on in [28]. The latter allows one party (Bob) to retrieve one of two possible pieces of information from another party (Alice), such that Bob does not gain any knowledge on the piece of data he has not retrieved nor Alice establishes which of the two data items she holds has been transferred to Bob. The 1-of-2 OT has been later generalised to k-of-n OT [29]. Oblivious transfer protocols can help realise a number of important cryptographic applications including zero-knowledge proofs and bit-commitment schemes.

This section explains how PUFs can be used to construct an oblivious transfer protocol using the 1-of-2 OT scheme proposed in [25].

The protocol described below requires a PUF with a large number of CRPs and an authenticated channel; it is run between two players; a sender (Alice) and a receiver (Bob).

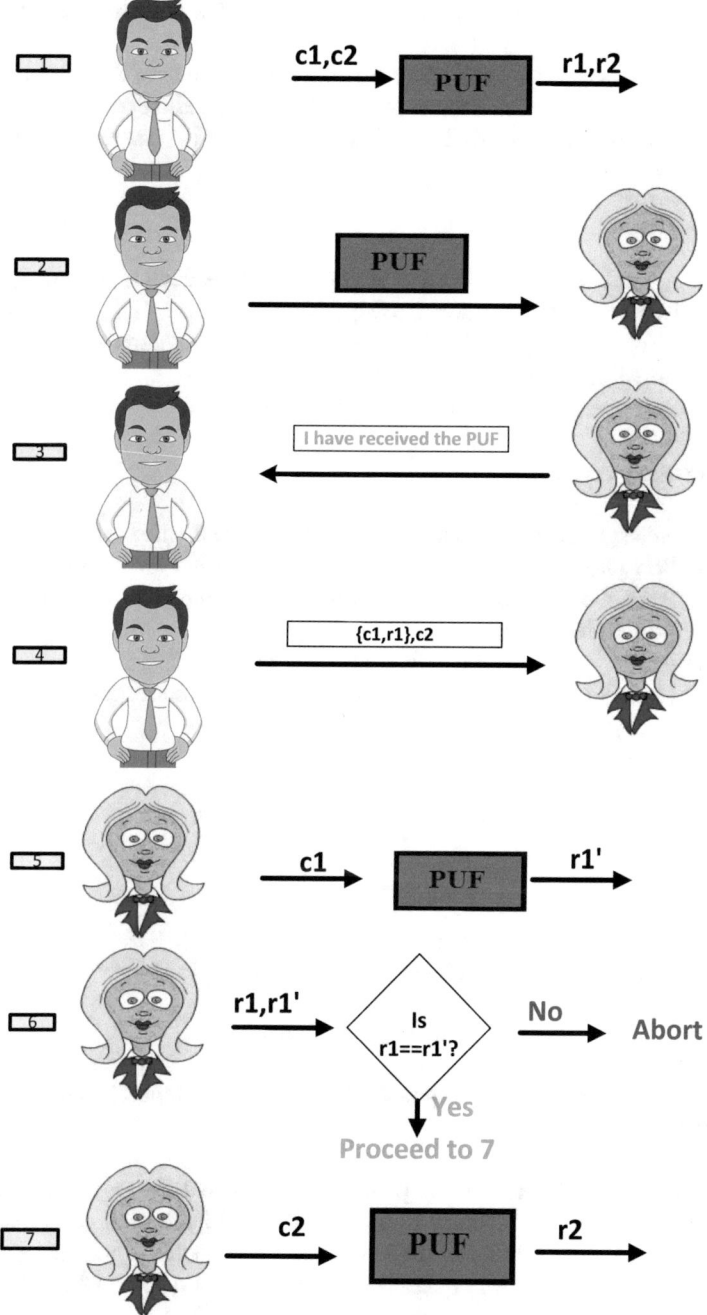

Fig. 6.7 A PUF-based key exchange protocol

At the start, Alice holds two secrets $b_0, b_1 \in \{0,1\}^\gamma$. And Bob makes a secret choice $s \in \{0,1\}$.

After the protocol is executed, Bob will have learned one of Alice's two secrets (b_s), without gaining any knowledge on the other item, and Alice will have learned nothing about Bob's choice (s). The protocol consists of two stages as explained in detail below:

(a) **Setup Phase**

  (1) A receiver (Bob) applies a random set of challenges $(c_0, c_1 \ldots, c_k)$ to a PUF and collects their corresponding responses $((r_0, r_1 \ldots, r_k)$, where $r_0, r_1 \ldots, r_k \in \{0,1\}^\gamma$. These challenge/response pairs (CRPs) are then stored in secure database (DB).
  (2) Bob gives the PUF to the sender (Alice).

(b) **Execution Phase**

  (1) Alice generates two random values (x_0, x_1) and Alice sends them to Bob.
  (2) Bob picks a challenge/response pair (c, r) from the database (DB) and sends the value $v = (c \oplus x_s)$ to Alice, For simplicity, let us assume that Bob secret choice was $(s = 0)$.
  (3) Alice applies the following two challenges to the PUF $\{c_0 = v \oplus x_0, c1 = v \oplus x_1\}$ and records the corresponding responses $\{r_0, r_1\}$.
  In this case

$$c_0 = v \oplus x_0 = c \oplus x_s \oplus x_0 = c \oplus x_0 \oplus x_0 = c$$

$$c_1 = v \oplus x_1 = c \oplus x_1 \oplus x_0$$

  (4) Alice sends $(r_0 \oplus b_0)$ and $(r_1 \oplus b_1)$ to Bob.
  (5) Bob can now deduce his chosen secret b_s as follows:

$$b_{s=0} = r_0 \oplus b_0 \oplus r = b_0$$

Note $r_0 = r$ as they are both responses to the same challnge c). The protocol is illustrated in Fig. 6.8.

It should be noted here that the protocol assumes that the responses, generated by Alice in step b-4, are identical to those that would have been generated in step a-1. This means the reliability of the PUF responses should be guaranteed; ideally, the PUF should have a 100% reliability metrics (please refer to Chap. 4 for more information on reliability enhancement techniques for PUFs).

The above protocol also assumes it is very unlikely that Bob can learn anything about the other secret (b_1), because with an overwhelming probability, he would not have measured (c_2, r_2).

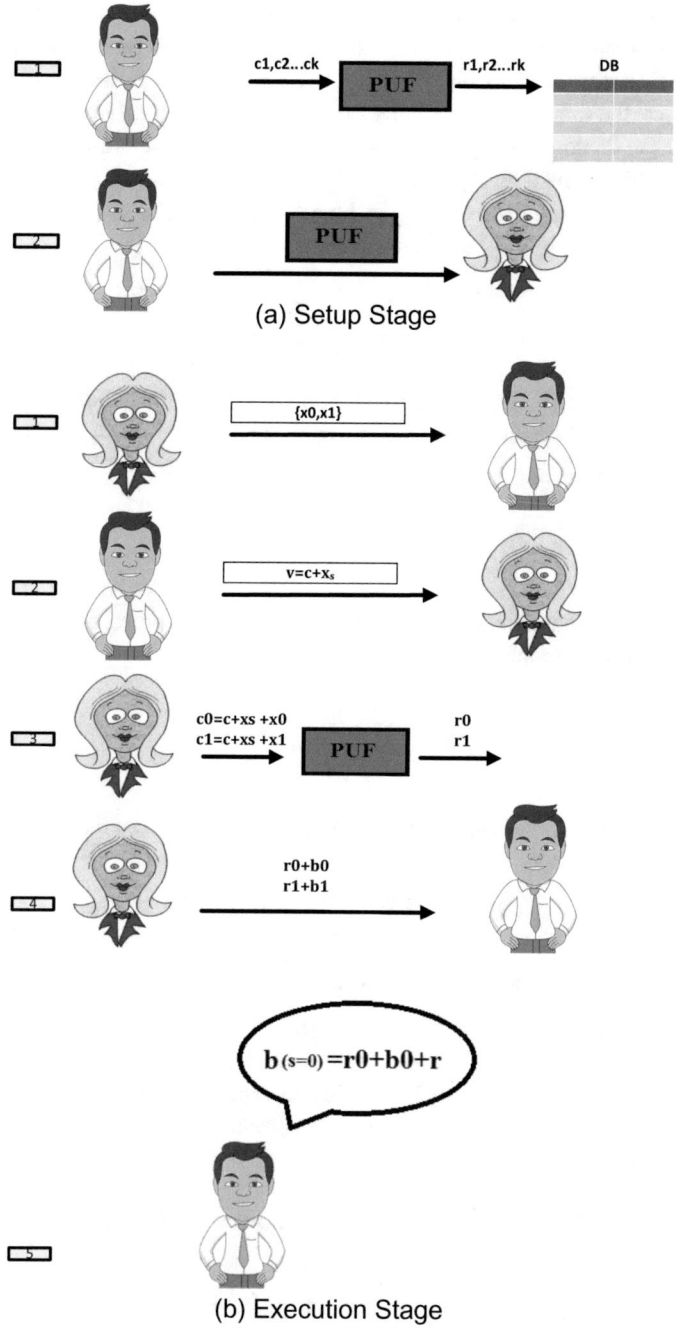

Fig. 6.8 A PUF-based 1-of-2 oblivious transfer protocol

To demonstrate this point, let us assume the PUF has (n) possible challenge/response pairs, out of which Bob collects a subset of size (k) in step a-1.

The probability of Bob obtaining (c_2), denoted as P, is the product of two probabilities: the first is the probability of Bob recording (c_2, r_2) in step a-1, denoted as P_1, and the second is the probability of Bob guessing c_2 assuming it exists in his database (DB), denoted as $P_2 = \frac{1}{k}$.

To compute P_1, the total number of possible subsets of size (k) in a set of size (n) is calculated using the binomial coefficient as follows:

$$\binom{n}{k} = \frac{n!}{(n-k)!k!} \tag{6.6}$$

In the same manner, it possible to compute the number of subsets which can be formed using all the challenges excluding c_2 as follows (assuming all challenges are equally probable)

$$\binom{n-1}{k} = \frac{(n-1)!}{(n-1-k)!k!} \tag{6.7}$$

The above two equations allow the computation of the probability that Bob may have recorded c_2 by chance, as given below:

$$P_1 = \frac{\frac{(n-1)!}{(n-1-k)!k!}}{\frac{n!}{(n-k)!k!}} = \frac{n-k}{n} \tag{6.8}$$

This means the probability of Bob obtaining c_2

$$P = P_1 \times P_2 = \frac{n-k}{n \times k} = \frac{1-k/n}{k} \tag{6.9}$$

Let us give a numerical example on how to compute P.

Example 6.3 Calculate the probability of Bob guessing c_2, hence computing b_1 as result of the execution of the 1-of-2 OT protocol described above. It is assumed that the PUF used has a total of $n = 2^{32}$ challenge/response pairs, out of which $k = 2^{20}$ is recorded by Bob in the setup stage of the protocol.

Solution:
Using Eq. (6.9), this probability can be computed as follows:

$$P = \frac{n-k}{n \times k} = \frac{2^{32} - 2^{20}}{2^{32+20}} = 9.53 \times 10^{-7}$$

6.5.4 Bit-Commitment (BC) Schemes

A commitment scheme is a cryptographic protocol which allows one party (referred to as the committer) to commit to a chosen value while keeping it secret from another party (referred to as the receiver). It consists of two stages: commitment and reveal.

Commitment schemes are often used to claim knowledge of some information without revealing it. For example, in a puzzle competition, a participant may publish a commitment to the solution of a puzzle so that if challenged at a later time, the participant can prove that they knew the solution of the puzzle at the time at which they published the commitment.

The basic working principles of this scheme can be explained as follows: consider a committer called 'Bob' and a receiver called Alice. At the commitment stage, Bob puts a message in a box and locks it; he then gives it to Alice and keeps the key. At the reveal stage, Bob proves his commitment by giving Alice the key to the box.

This scheme has a number of important applications such as verifiable secret sharing [30]. Another use is secure billing protocols [31, 32], wherein a consumer can prove to a utility provider his commitment to energy costs, without revealing the actual value of the metre readings, which can be beneficial in protecting the consumer's usage details (when and how they consume their energy).

This protocol can be constructed using the PUF-based oblivious transfer protocol described above; this is achieved by reversing the roles of the sender and receiver [25]. In other words, Bob the OT-receiver acts as BC-committer, and Alice the OT-sender acts as the BC-receiver. The procedure of the protocol is as follows:

(a) *Commitment Stage*

   (1) The BC-sender (Bob) acts an OT-receiver and uses his secret choice (s), where $s \in \{0, 1\}$, as an input to the PUF-OT protocol described above.
   (2) The BC-receiver acts as an OT-sender and used his secret values (b_0, b_1), where $b_0, b_1 \in \{0, 1\}^\gamma$, as inputs to the OT protocol.
   (3) The OT protocol is then run which allows Bob to learn one of Alice's secrets (i.e. b_s).

(b) *Reveal Stage*

   (1) Bob sends Alice the binary string (s, b_s) (the fact that Bob was able to compute b_s proves his previous commitment to the secret choice s).

6.6   Remote Secure Sensors

6.6.1   Motivation

The general concept of a wireless sensor network is that of deploying many small, inconspicuous, self-contained sensor nodes into an environment to collect and transmit information, and possibly provide localized actuation. Potential uses for such networks include medical applications; structural monitoring of buildings; status monitoring of machinery; environmental monitoring; military tracking; security; wearable computing; aircraft engine monitoring and personal tracking and recovery systems [33–35].

Security is an important requirement in some of these applications such as sensors used for remote health monitoring systems that handle private patients' data. Other security-sensitive applications include nuclear and chemical material tracking systems. Existing solutions for secure remote sensing rely on the use of a cryptographic block that performs data encryption/authentication tasks as shown in Fig. 6.9 [36].

This solution, however, suffers from two shortcomings. First, the data output of the sensing element is not protected before it enters the cryptographic block, so in principles, an adversary who can gain physical access to the sensor can read/manipulate the sensor measurements using invasive physical attacks (e.g. directly probing the internal signals) or non-invasive side channel analysis [37]. The second disadvantage of the existing design is its reliance on the use of classic cryptographic primitives such as symmetric ciphers or hash functions, which may be prohibitively expensive or not even feasible in the case of resources-constrained sensors [12].

The technology of physically unclonable functions provides an alternative approach for designing secure sensors, which does not require a separate cryptographic module. Several PUF-based secure sensing schemes have already been proposed in the literature [15, 38, 39]. The basic working principles of a PUF-based secure sensor will be explained in the following section.

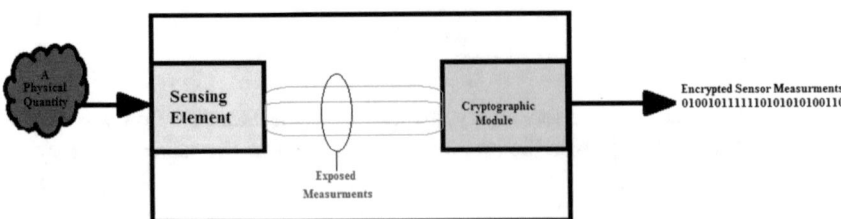

Fig. 6.9 A typical architecture for a secure sensor design

6.6.2   Principles of PUF-Based Secure Sensors

PUF circuits are susceptible to environment parameter variations such as temperature or supply voltage, which effectively means that the response of a PUF depends on both the applied challenges and on the ambient parameters.

Although the unreliability of a PUF design may hinder its use in applications such as key generation and authentication schemes, it makes it an attractive choice for designing secure sensors.

The essence of this approach is illustrated in Fig. 6.10, first the physical quantity is translated to an electrical signal using a transducer. This signal is then used to modulate the supply voltage of a PUF circuit, which leads to a change in some of its response bits. The number of changing bits is dependent on the magnitude of the modulating signal, hence the physical quantity being measured.

This architecture can be used to measure any physical quantity as long as the latter can be translated into an electrical signal.

This technique assumes that there is an enrolment stage before the sensor is deployed, during which the challenge/response behaviour of the PUF is characterised under different ambient conditions (i.e. voltage supply, temperature). This information is subsequently used to translate the output of the PUF sensor into a meaningful measurement data, similarly to the decryption stage in classic cryptography.

This scheme does not need a data encryption stage to protect against an eavesdropping adversary, because only the parties who have access to the PUF or its characterisation data can map the obtained responses to meaningful data measurements.

To illustrate how such a concept can be implemented in practice, consider a wireless sensor network that has one central server and multiple sensor nodes.

Consider a PUF circuit which has N inputs and M outputs, and can operate at different voltage levels:

$$\{v_{min}, v_{min} + \Delta v, \ldots, v_{min} + (k-1)\Delta v\}$$

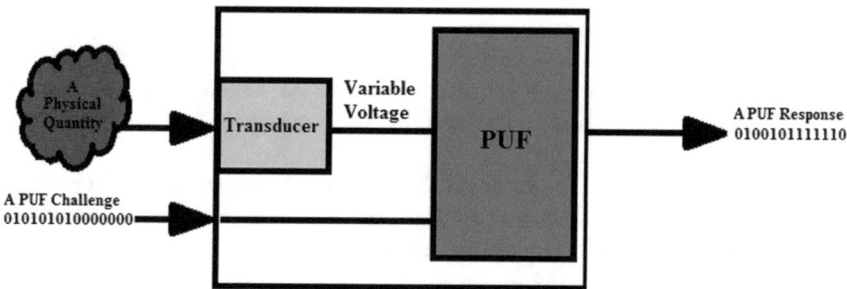

Fig. 6.10   A PUF-based architecture for a secure sensor design

Table 6.5 Example entries of PUF sensor database

C	v_{min}	$v_{min} + \Delta v$	...	$v_{min} + (k-1)\Delta v$
c_0	r_{00}	r_{01}		$r_{0(k-1)}$
	PQ_{00}	PQ_{01}		$PQ_{0(k-1)}$
c_1	r_{10}	r_{11}		$r_{0(k-1)}$
	PQ_{10}	PQ_{11}		$PQ_{1(k-1)}$

where

v_{min} is the minimum supply voltage at which the PUF circuit can still generate an output.

$v_{min} + (k-1)\Delta v$ is the maximum supply voltage that can be applied to the PUF circuit without damaging it.

Δv is the minimum step change in the supply voltage needed to cause one-bit change in the PUF response.

It is worth noting here that the minimum/maximum values of the voltage supply are a function of the implementation technology; therefore, one should consider technologies which allow a large range of operating voltages. For simplicity, we assume that changes in the PUF responses due to ambient temperature are negligible. The implementation procedure can now be outlined as follows:

(a) *Enrolment*

    (1) For each PUF, a number of challenges are applied and their corresponding responses are measured.

    (2) The above experiment is repeated for a (k) different supply voltages.

    (3) The server needs also to establish the mapping between the physical quantity to be measured and the voltage supply applied to the PUF.

    (4) For each PUF sensor, the central server creates a database that includes the challenges, their responses at different supply voltages and the values of the physical quantity corresponding to each voltage level. Exemplar characterization data are shown in Table 6.5, which include the entries for two challenges (c_0, c_1) applied to a PUF instance.

(b) *Sensing*

    (1) The server transmits a challenge (c) from its database to a PUF sensor.

    (2) The PUF generates a response (r), which is a function of the received challenge and the physical quantity being measured, and sends it back to the server.

    (3) The server uses the received response to find the corresponding supply voltage, which then mapped to the physical quantity being measured (air pressure, chemical substance, etc.).

    (4) The server may delete the used challenge in order to protect against replay attacks.

        This scheme assumes that only the server has access to PUF characterization data, so that responses can be transmitted without encryption.

6.6.3   PUF Design Requirements

This section discusses the qualities a PUF design should have to be suitable for secure sensing applications.

Let us consider the architecture shown in Fig. 6.10, it assumes that the change of the PUF response to a certain challenge only reflects the purposely induced fluctuations of the supply voltage (Δv), hence the physical quantity being measured (PQ). Such an assumption, however, ignores the fact that the PUF circuitry may still be susceptible to natural variations in ambient parameters (e.g. temperature, supply voltage) and to other sources of temporal noise, as discussed in Chap. 3. Therefore, in order to ensure a PUF design is suitable, it needs to be made sensitive to the purposely induced voltage supply change (Δv), but resilient to all other forms of noise. In order to evaluate the suitability of a PUF design for a secure sensor application, the following two metrics will be used:

1. **The InterPQ distance** $\left(HD_{InterPQ}\right)$ which refers to the Hamming distance between two PUF responses to the same challenge evaluated at two supply voltages which differ by (Δv) but under the same noise conditions.
2. **The IntraPQ distance** $\left(HD_{IntraPQ}\right)$ which refers to the Hamming distance between two PUF responses to the same challenge evaluated at the same nominal supply voltage but under different noise conditions.

Based on the above metrics, one can deduce the minimum requirements a PUF needs to satisfy to be suitable for a sensor application as follows:

Definition 6.1 *Sensing Usability Condition*

   A PUF circuit that has N inputs and M outputs, and can operate under k different voltage levels $(v_1, v_2 \ldots v_k)$, is considered to be suitable for a sensor that can measure k distinct value of a physical quantity (PQ) if it has at least one challenge (c) which produces a set of responses at different supply voltages.

   $R_v = \{r_{v1}, r_{v2} \ldots r_{vk}\}$ such as

$$\forall r_{v1}, r_{v2} \in R_v$$

$$min(HD_{InterPQ}) > max(HD_{IntraPQ}) \qquad (6.10)$$

$$min(HD_{InterPQ}) > 1 \qquad (6.11)$$

This condition basically means that the minimum Hamming distance between PUF responses to the same challenge under different supply voltages should be more than one, it should also be larger than the maximum number of response bit flips caused by temporal noise conditions.

Another desirable quality in a PUF sensor is uniqueness, so that if one device is compromised (e.g. its behaviour is learned by an adversary), the others remain secure. Ideally, the uniqueness should be 50%.

Let us now give a numerical example to illustrate how the sensing usability condition can be verified.

Example 6.4 Consider a PUF circuit that has two inputs and six outputs, wherein its challenge/response pairs at various supply voltages is shown in Table 6.6. Explain whether or not this circuit can be employed as a sensor capable of differentiating between five distinct levels of the supply voltage. It is assumed in here that $\Delta v = 0.1v$, and the maximum change of response bits due to temporal noise is 2.

Solution:

To establish the suitability of the PUF for sensor applications, we refer to Definition 6.1.

First, we need to compute the Hamming distance between each pair of responses for each challenge.

For $c = 00$, we have

$\min(HD_{InterPQ}(r_{0.8}, r_{0.9})) = 0 < 1$, which means the PUF is not sensitive to the minimum step change of the supply voltage for this challenge; therefore, we need to try other challenges.

For $c = 01$, we have

$$HD_{InterPQ}(r_{0.8}, r_{0.9}) = 2$$

$$HD_{InterPQ}(r_{0.8}, r_1) = 2$$

$$HD_{InterPQ}(r_{0.8}, r_{1.1}) = 2$$

$$HD_{InterPQ}(r_{0.8}, r_{1.2}) = 2$$

$$HD_{InterPQ}(r_{0.9}, r_1) = 4$$

Table 6.6 PUF challenge/response at various supply voltages

Challenges	Voltage supply (v)				
	0.8	0.9	1	1.1	1.2
00	001000	001000	001000	001010	001010
01	000000	110000	001100	000011	100001
10	111011	111000	111000	111111	111111
11	000111	000111	000111	000110	000111

$$HD_{InterPQ}(r_{0.9}, r_{1.1}) = 4$$

$$HD_{InterPQ}(r_{0.9}, r_{1.2}) = 2$$

$$HD_{InterPQ}(r_1, r_{1.1}) = 4$$

$$HD_{InterPQ}(r_1, r_{1.2}) = 2$$

$$HD_{InterPQ}(r_{1.1}, r_{1.2}) = 2$$

The above calculation indicates $c = 01$ satisfies both equations stated in Definition 6.1, hence the sensing usability condition.

Therefore, this circuit can be used as sensor using $c = 01$.

6.6.4   Design Case Study: Arbiter PUF Sensors

This section provides an exemplar case study of a PUF sensor design using a 32-bit arbiter PUF.

(a) *Design Specification*

There are five main requirements that need to be defined at this stage:

(1) The minimum step change in the supply voltage Δv which is a function of the resolution of the transcoder output.
(2) The number of distinct levels of the physical quantity to be measured (k).
(3) The upper bound of implementation costs (area, energy, etc.).
(4) False acceptance rates (FAR): the probability of generating a valid response of the wrong PQ value, an example of such a case is when the PUF generates the same response for two different PQ values.
(5) False rejection rate (FRR) is the probability of generating an invalid response for a genuine PQ value, this is typically due to external noise.

FRR and FAR can be estimated using the $HD_{InterPQ}$ and $HD_{IntraPQ}$ as follows [15]:

$$FRR = 1 - \sum_{i=0}^{m} \binom{M}{i} (\hat{p}_{IntraPQ})^i (1 - \hat{p}_{IntraPQ})^{M-i} \qquad (6.12)$$

$$FAR = \sum_{i=0}^{m} \binom{M}{i} (\hat{p}_{InterPQ})^i (1 - \hat{p}_{InterPQ})^{M-i} \qquad (6.13)$$

wherein

M is the number of PUF output bits (i.e. the bit length of its responses per challenge).

m is the maximum number of noise-induced errors of a PUF response which does not cause a false rejection, in other words, the allowed error margin.

$\hat{p}_{IntraPQ}, \hat{p}_{IntrePQ}$ are the binomial probability estimators of $HD_{IntraPQ}$ and $HD_{InterPQ}$, respectively.

It is worth noting here that the above equations assume that $HD_{IntraPQ}$ and $HD_{InterPQ}$ have binomial distributions.

Ideally, both FRR and FAR need to be minimised; however, Eqs. (6.12) and (6.13) indicate that in order to minimise FRR, m need to be as large as possible so the bit flips caused by temporal noise do not lead to rejecting a genuine response. On the other hand, increasing m can aggravate FAR, because it increases the probability of obtaining the same response from different physical quantities. Therefore, in practice, m is chosen to balance FRR and FAR, in other words, to make them equal. This value of m is referred to as the equal error threshold (m_{EER}). For discreet distributions, it may not be possible to find such a value, instead m_{EER} is computed as follows:

$$m_{EER} = \mathrm{argmin}_m \{\max\{FAR(m), FRR(m)\}\} \qquad (6.14)$$

And the equal error rate, in this case, will be

$$EER = \max\{FAR(m), FRR(m)\} \qquad (6.15)$$

In this design study, the design requirements are as follows:

$$\Delta v = 0.15V, k = 3 \ and \ EER < 10^{-4}.$$

In addition, the silicon area needs to be minimised.

(b) **PUF Implementation and Optimization**

The goal of this stage is to find the best implementation of the PUF such as all design specifications are met. The arbiter PUF has been chosen due to the relative ease of its design. Figure 6.11 shows an illustration of an arbiter PUF with 32 inputs (N = 32) and one output (M = 1), it consists of two delay paths that have the same layout, hence nominal latency. The selection inputs (i.e. the challenges) create different configurations of these two paths, for each challenge, the delay of the signal (X) through the chosen two paths is compared to produce a response (Y).

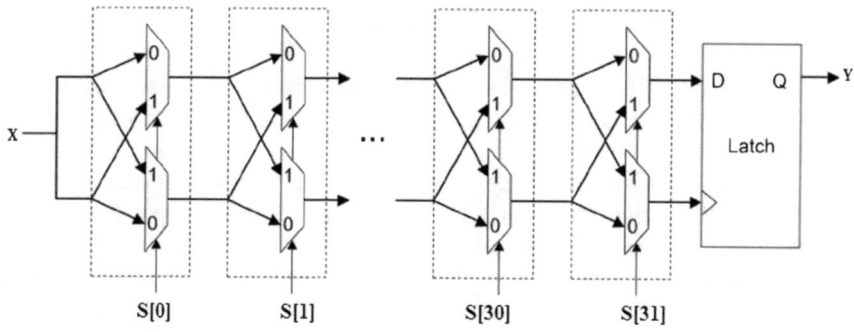

Fig. 6.11 A 32-stage arbiter PUF

The choice of implementation technology is typically up to the designer, but it can also be defined at the specification stage, in this case, 65 nm CMOS technology library is used; it has a nominal supply voltage of 1.2 V. It is also assumed that there are only two noise sources: temperature variations (−40 to 85 °C) and power supply fluctuations of 5%.

The first step in the implementation is to specify the three voltage levels to be used, in this case $\Delta v = 0.15$, so we chose $\{0.9, 1.05, 1.2\}$ which are within the operational range of the chosen technology.

The second step is to find the appropriate number of PUF inputs and outputs (N, M), to do that one needs to compute *FAR* and *FRR* for different design configurations.

To increase M, one can use multiple single-response arbiter PUFs, and to increase N, one can use more delay stages as shown in Fig. 6.10. This case study considers three configurations ($N = 32$, $M = 32, 64, 128$). In each, the following experiment is carried out:

(1) 2000 challenge/response pairs are randomly selected.
(2) For each challenge, both the average values of $HD_{IntraPQ}$ and $HD_{IntrePQ}$ of its responses are evaluated.
(3) The results from the previous step are used to computed *FAR* and *FRR*.
(4) The experiment is repeated for different m (allowed error margin) in order to find the m_{EER} in each case.

The results outlined in Table 6.7 indicate that the best solution out of the considered three options is $N = 32, M = 64$.

Table 6.7 The impact of response bit length on the FAR/FRR of an arbiter PUF-based sensor

N	M	m_{EER}	FAR	FAR
32	32	7	5.3×10^{-4}	5.5×10^{-4}
32	64	7	1.01×10^{-5}	1.05×10^{-5}
32	128	8	8.6×10^{-6}	9.7×10^{-6}

6.7 Anti-counterfeiting Techniques

PUF technology can be employed to limit overproduction of integrated circuits by malicious factories, which is causing significant financial losses to design houses every year [40].

This is achieved by embedding each chip with a PUF circuit and a locking mechanism during the design stage. At the post-fabrication stage, the foundry applies challenges chosen by the designer to each PUF and sends the corresponding responses back to the design house. The designer then authenticates each device and computes a passkey using a response to a known challenge. The passkey is then sent back to the foundry to activate the chip for testing purposes. This process is sometimes referred to as active hardware metering, and it gives designers post-fabrication control over their designed chips, so that only authenticated chips can be used [41].

In order to explain how this approach works, we are going to give an example using a sequential digital design based on Mealy state machine. The detailed procedure, in this case, consists of the following steps:

1. The design house embeds a lock mechanism at the register transfer level (RTL) level. There a number of approaches to design such a lock; in the case of sequential digital circuits, this can be done by adding non-functional states to the original state machines. Take, for example, the state machine shown in Fig. 6.12, the grey states represent the actual design, and the black states represent the redundant non-functional states. These additional states serve two purposes: first they obfuscate the original functionality of the design, hence making it harder to understand and copy. Second, they can be used to lock the design in a non-functional state, such that only a designer or an authorised party has the knowledge to bring the design into a functional state by applying the correct sequence of inputs.

2. The design house embeds a PUF in each chip as shown in Fig. 6.13, the outputs of the PUF are used to initialize the internal flip-flops of the design (i.e. place the design in an initial state). The output of the PUF is random, and therefore, all the states in the design can become an initial state with equal probabilities; therefore, in order to increase the probability that a design starts in a non-functional state (e.g. a black state in Fig. 6.12), the number of the redundant states should be significantly larger than the number of functional states (e.g. the grey states in Fig. 6.12). This does mean there will be a large area overhead, but this extra cost is the price to pay to ensure post-fabrication control and limit illegal chip overproduction.

   In addition to the extra flip-flops incurred by adding the extra states, the increase in area is caused by the additional multiplexer and the PUF circuits. However,

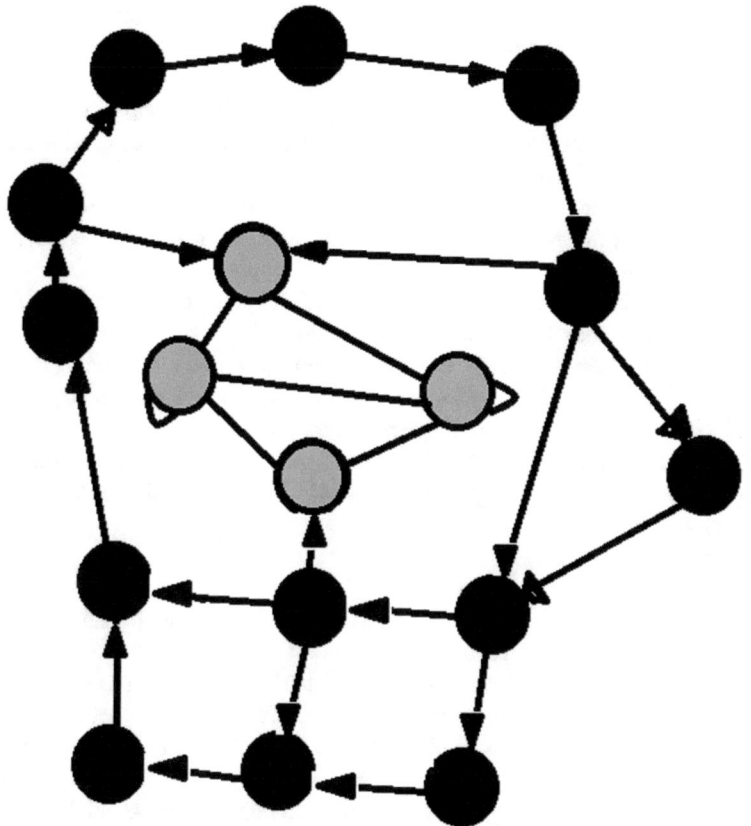

Fig. 6.12 Sequential 1 design obfuscation using redundant states

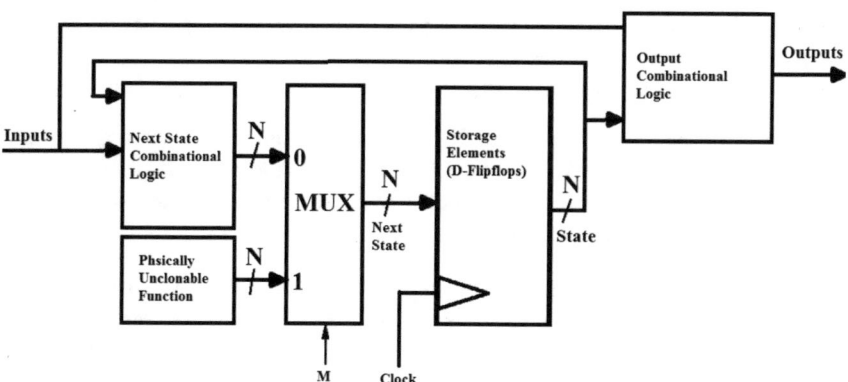

Fig. 6.13 Circuit architecture of a PUF-initialised synchronous sequential logic based on mealy state machine

this can be reduced by reusing some existing functional blocks such as the multiplexers used for on-chip testing purposes (e.g. scan path), which each chip normally has.

3. The design house sends the post layout files (e.g. GDSI files) to the fabrication facilities, these files are typically in a non-readable format, along with a specific challenge for the PUF.

4. Once the chips are fabricated, the manufacture will read-out the response of the PUF to the designer challenge from each device and send these back to the design house.

5. The design house computes a key to unlock each chip based on its PUF response and sends it back to the manufacturer.

6. For each chip (see Fig. 6.13), the manufacture powers-up the device, applies the received challenge to the PUF, set (M = 1) and applies one clock cycle. This will set the design in the PUF-generated initial state.

7. The manufacturers then sets (M = 0) and then applies the key to the primary inputs and clock the design. This should drive the design into one of the functional states.

8. The chip is now ready for testing.

For chip unlocking, the above procedure needs only to be applied once to provide protection against chip overproduction; therefore, the designer needs to ensure that the design can never go back to a locked state after it has been unlocked. One way to achieve this is to store the uncloaking key for each device in a non-volatile memory on chip, such that each time the device is powered-up, the internal state is automatically initialised by the PUF to lock the design; then, the stored key is automatically applied to unlock it. In this case, there will be no need to protect encrypt or protect the unlocking key for each chip, this is because each device will have its own key which is useless to use on other devices.

It is worth noting here that the above approach assumes the PUF is able to produce the same response consistently, which implies there may be a need for additional reliability-enchantment techniques as discussed in Chap. 4 such as error correction codes and stable-bit selection.

We are now going to give an illustrative example using a sequential circuit.

Example 6.5 The state machine in Fig. 6.14 represents a synchronous sequence detector, and it has one output and one input in addition to the clock input. It works as follows: the output will generate a pulse every time the following 4-bit sequence of input is applied (0111) serially on its input.

(1) How many flip-flops you need to represent all the state in this design?

(2) Devise a new state machine which obfuscates the original functionality of this design, such as the total number of additional flip-flop needed does not exceed 2.

Fig. 6.14 A state machine for a 4-bit sequence detector

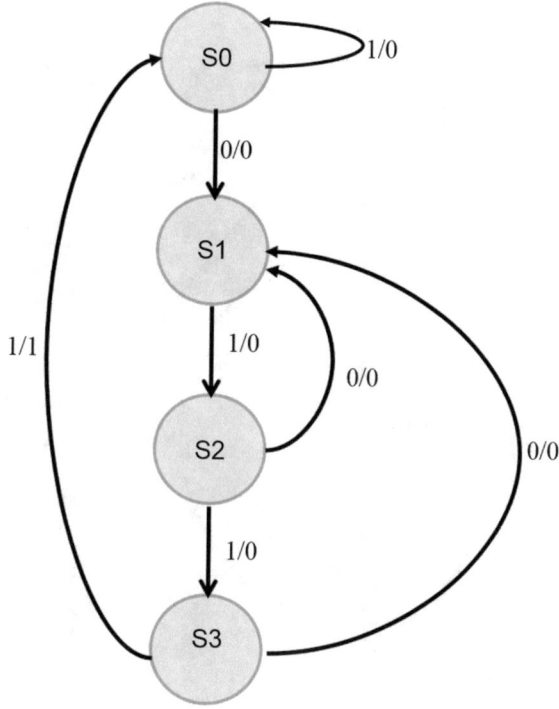

Solution:

(1) This design has four states; therefore, two flip-flops are needed to represent all the states.

(2) Figure 6.15 shows an exemplar obfuscated design for the same circuit, we have added a total of 12 redundant states, so in this case, four flip-flops will be needed to represent the internal states. It is worth noting that in this solution the output of the circuit remains low if the design is in one of the non-functional states (S4–S15). In this, case the probability of the design starting accidentally in a functional state is 25%, which is still very high. To make the design more secure, one needs to add more flip-flops.

Example 6.6 The obfuscated state machine in Fig. 6.15 is adopted to develop circuit architecture for a PUF-initialised synchronous sequence detector as shown in Fig. 6.13. Binary coding is used to represent each state (e.g. S1 is coded as 0001, S2 as 0011, etc.)

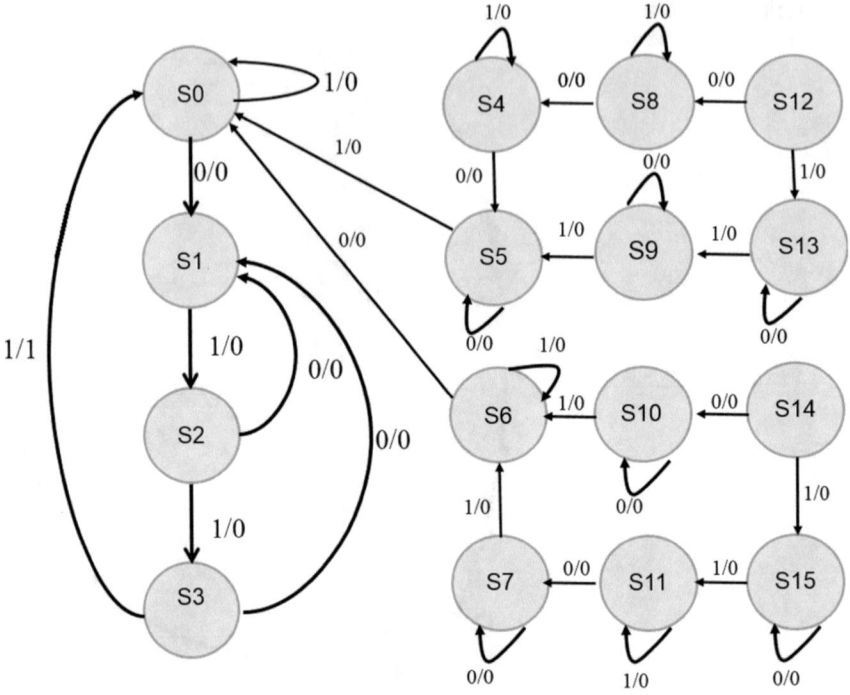

Fig. 6.15 A state machine for an obfuscated 4-bit sequence detector

(1) What is the required length of the PUF's response?
(2) Compute the unlocking key for the following devices, given their associated
     PUF responses.

   Device 1(1111), Devise 2(0111), Devise 3(1110)
   Solution:

(1) The PUF needs to have a 4-bit response length in order to initialise the
     state-holding flip-flops.
(2) For Device 1: the PUF response is 1111, which means the design is going to
     be initialised to the non-functional state (S15). Looking at the state machine in
     Fig. 6.15, the following sequence of input needs to be applied from left to
     right 1010, this can be used as the unlocking key.

   To clarify, the first bit of this sequence brings the design from S15 to S11, the
second bit of this sequence brings the design from S11 to S7, the third bit of this
sequence brings the design from S7 to S6, and the last bit of this sequence brings
the design from S6 to S0.
   The unlocking keys for the other two devices can be computed in the same
manner.

6.8   Tamper-Proof Design

PUF technology can be used to design anti-tamper hardware; this is particularly important for applications wherein devices are left unprotected; in such cases, a coating-based design can be used. The latter is placed on the top layer of an integrated circuit (IC). One implementation of this approach is to lay out a network of metal wires as a comb shape; the latter produces a capacitor unique for each device because of fabrication-induced variations in the placement size and dielectric strength. This capacitor is used to obtain a unique identifier for the device carrying the Coating PUF. When an attacker tries to tamper with such a device, he may remove a part of the coating, which leads to a change in the capacitor value, hence destroying the unique identifier of the device. An example of PUF-based anti-tamper integrated circuits is shown in [42].

6.9   Conclusions

A summary of important lessons is provided below:

- Physically, unclonable functions can be employed for on-chip generation of root keys. This provides an alternative approach to the existing key provisioning process, which consists of an off-chip generation followed by on-chip memory writing. The PUF approach provides a better protection of root keys against physical attacks on hardware; it also enhances the security of the key provision process because the device manufacturers will no longer be required to handle key generation/injection. However, PUF-based key generation schemes incur large implementation costs because of the need for extra error correction logic to ensure stable key generation in the field. There is ongoing research on how to generate a stable key from a PUF device in the presence of temporal noise and with the minimum possible costs.
- PUF technology can also be used to authenticate hardware devices (e.g. smart cards, e-passports) by exploiting the distinctive and unique challenge/response behaviour each PUF instance exhibits when implemented on different chips. Such usage makes it feasible to design entity authentication protocols suitable for resources-constrained systems such as RFIDs. The majority of proposed PUF-based authentication schemes require that a verifier characterises the challenge/response behaviour of each PUF, and creates a database which stores this information. This process may not efficient if the number of devices to be authenticated is large. The ongoing research in this area is focused on increasing the scalability of the characterisation process.
- PUF technology is also employed to design low-cost secure sensors; this is achieved by exploiting the change of the challenge/response behaviour of a PUF circuit due to fluctuations in power supply. In this case, a transducer converts a physical quantity (pressure, chemical material, etc.) to an electrical signal, the

latter modulates the power supply of the PUF circuit. Such usage requires the PUF design to be capable of functioning correctly in a range of supply voltages and demonstrating distinctively measurable challenge/response behaviour in each case. Ongoing work in this area focuses on developing PUF architectures which can provide correct measurement in spite of temporal noise and normal fluctuation in operating conditions, in other words: how to design a PUF circuit that is sensitive to purposely induced power supply fluctuations and resilient to temperature and voltage variations?

- PUF technology is increasingly adopted in advanced security protocols such as key exchange schemes and oblivious transfer techniques.
- Other PUF applications include Anti-Counterfeiting ICs, intellectual property protection, and software licensing.
- In summary, PUF technology has a great commercial potential, but there are still numerous design challenges need overcoming before such potential can be converted into actual products.

6.10 Problems

(1) A single PUF is used to generate a 128-bit encryption key, wherein a BCH code (n, k, t) = (15, 7, 2) is employed to generate the helper data.

    (a) How many raw PUF response bits are needed assuming the required secrecy rate is 0.9?

    (b) How many challenges would you need to apply to generate those raw bits given the PUF has a 32-bit response length?

(2) The challenge/response behaviour of a PUF is given in Table 6.8. The design is used in the authentication protocol depicted in Fig. 6.4. The PUF is expected to operate under six different environment conditions (i.e. degree of power supply fluctuations and range of ambient temperature). The nominal conditions at which the PUF was enrolled at are (25 °C, V = 1v) It is assumed

Table 6.8 PUF challenge/response behaviour under different environment conditions

Environment conditions	Challenges			
	$c_0 = 00$	$c_1 = 01$	$c_2 = 10$	$c_3 = 11$
$T = 25\,°C, V_{dd} = 1v$	00000000	00000111	00111000	11111111
$T = 75\,°C, V_{dd} = 1v$	00000000	00000111	00110000	11111000
$T = 25\,°C, V_{dd} = 1.2v$	00000001	10000111	00100000	01111000
$T = 75\,°C, V_{dd} = 1.2v$	00000001	11000111	00000000	01111000
$T = 25\,°C, V_{dd} = 0.8v$	00000011	11000111	00000000	01111000
$T = 75\,°C, V_{dd} = 0.8v$	00000011	11000111	00000001	01111011

all challenges listed in Table 6.8 have equal probabilities, and all the listed environment conditions are equally probable.

(a) What is the best threshold value that minimises the probability of a denial of service?
(b) What is the best threshold value that minimises the probability of a forged PUF being authenticated?
(c) What is the authentication threshold value (t) that produces the minimum equal error rate (EER)?

(3) A PUF is used to construct a 1-of-2 OT protocol described in Fig. 6.8 Compute the maximum number of challenge/response pairs that can rerecord by Bob at the setup stage, such that the probability of him guessing b_1 as a result of one execution of protocol is less than 10^{-8}, given the total number of challenge/response pairs of 2^{64}.

(4) Consider a PUF circuit that has two inputs and eight outputs, and its challenge/response behaviour is given in Table 6.9. Explain whether or not this circuit can be used to design a secure sensor such that shown in Fig. 6.10. It is assumed in here that $\Delta v = 0.1v$, and the maximum change of response bits due to temporal noise is 1.

(5) A 256-bit key generation scheme is to be constructed using a PUF design, whose reliability in the expected operating environment is given as 91%. The PUF has a 16-bit input and 32-bit output and an area of 123 um^2. Assume that code-offset scheme is used for helper data generation as shown in 6.3.4.

(a) What is the minimum number of raw PUF bits needed to obtain this key assuming the secrecy rate is 0.85?
(b) Table 6.10 gives two possible error correction codes with their associated area. Compute the number of helper data needed by each code if it is to be used as part of the key generation scheme.
(c) Which of the codes listed in Table 6.10 lead to a smaller area overhead of the overall key generation scheme?

Table 6.9 PUF challenge/response at various supply voltages

Challenges	Voltage supply (v)				
	0.8	0.9	1	1.1	1.2
00	00000010	00000011	00000000	00000001	10000001
01	11000110	11000111	00000111	10000111	00000111
10	11000110	11000111	00111000	00100000	10100000
11	01111000	01111000	11111111	01111000	01111000

Table 6.10 Error correction codes

Code	Codeword length (n)	Dataword length (k)	Number of c	Area overhead (um^2)
BCH (31,16,3)	31	16	3	344
BCH (63,30,6)	63	30	6	579

(6) Which of the following attacks are the most likely threat to the PUF-based oblivious transfer

    (a) Modelling attacks using machine learning algorithms.
    (b) Physical cloning attacks.
    (c) Side channel analysis.

(7) The false rejection/admission rates of a basic PUF authentication protocol (as shown in Sect. 6.4.3) are 1.2×10^{-6} and 2.3×10^{-6}, respectively. A code-offset scheme is employed to improve the reliability of the used PUF design. How does the use of this scheme affect the false rejection/admission rates?

(8) A ring oscillator PUF is used in a secure sensing application (as shown in Sect. 6.6.2), wherein the false rejection/admission rates are 3.5×10^{-6} and 4.9×10^{-6}, respectively.

    (a) What is the equal error rate in this case?
    (b) Arrange the following techniques in terms in their effectiveness in reducing the above error rate: (1) ageing acceleration, (2) secure sketch schemes and (3) stable response selecting.

(9) The state machine in Fig. 6.14 represents a synchronous sequence detector, and it has one output and one input in addition to the clock input. It works as follows: the output will generate a pulse every time the following 4-bit sequence of input is applied (0111) serially on its input.

    (a) Devise a new state machine which obfuscates the original functionality of this design, such that the total number of additional state-holding flip-flops needed is 3.
    (b) What is the probability of the design accidentally starting in a functional state?
    (c) A PUF with a 5-bit response length is used to initialise the design. Compute the unlocking key of the following devices based on your solution for question 9-b.

References

1. P. Stewin, I. Bystrov, Understanding DMA Malware, ed. by U. Flegel, E. Markatos, W. Robertson. Revised Selected Papers Detection of Intrusions and Malware, and Vulnerability Assessment: 9th International Conference, DIMVA 2012, Heraklion, Crete, Greece, July 26–27, 2012, (Berlin: Springer, 2013), pp. 21–41
2. S. Skorobogatov, in *Data remanence in flash memory devices*. Presented at the Proceedings of the 7th International Conference on Cryptographic Hardware and Embedded Systems, Edinburgh, UK, 2005
3. J.A. Halderman, S.D. Schoen, N. Heninger, W. Clarkson, W. Paul, J.A. Calandrino et al., Lest we remember: cold-boot attacks on encryption keys. Commun. ACM **52**, 91–98 (2009)
4. S.K. Mathew, S.K. Satpathy, M.A. Anders, H. Kaul, S.K. Hsu, A. Agarwal, et al., 16.2 A 0.19pJ/b PVT-variation-tolerant hybrid physically unclonable function circuit for 100% stable secure key generation in 22 nm CMOS in *2014 IEEE International Solid-State Circuits Conference Digest of Technical Papers (ISSCC)*, (2014), pp. 278–279
5. M.T. Rahman, F. Rahman, D. Forte, M. Tehranipoor, An aging-resistant RO-PUF for reliable key generation. IEEE Trans. Emerg. Top. Comput. **4**, 335–348 (2016)
6. Z. Paral, S. Devadas, in *Reliable and efficient PUF-based key generation using pattern matching*. 2011 IEEE International Symposium on Hardware-Oriented Security and Trust, 2011, pp. 128–133
7. R. Maes, *Physically unclonable functions: constructions, properties and applications* (Springer, Berlin, 2013)
8. Development and Education Board (2017), available: https://www.altera.com/solutions/partners/partner-profile/terasic-inc-/board/altera-de2-115-development-and-education-board.html
9. S.S.K.J. Guajardo, G.-J. Schrijen, P. Tuyls, in *FPGA intrinsic PUFs and their use for IP protection*. International Conference on Cryptographic Hardware and Embedded Systems, pp. 63–80, 2007
10. T. Ignatenko, G.J. Schrijen, B. Skoric, P. Tuyls, F. Willems, in *Estimating the secrecy-rate of physical unclonable functions with the context-tree weighting method*. 2006 IEEE International Symposium on Information Theory, 2006, pp. 499–503
11. Y. Dodis, L. Reyzin, A. Smith, in *Fuzzy extractors: how to generate strong keys from biometrics and other noisy data*, ed. by C. Cachin and J. L. Camenisch. Proceedings on Advances in Cryptology—EUROCRYPT 2004: International Conference on the Theory and Applications of Cryptographic Techniques, Interlaken, Switzerland, May 2–6, 2004, (Berlin: Springer, 2004), pp. 523–540
12. S. Satpathy, S. Mathew, V. Suresh, R. Krishnamurthy, in *Ultra-low energy security circuits for IoT applications*. 2016 IEEE 34th International Conference on Computer Design (ICCD), (2016), pp. 682–685
13. T. Xu, J. B. Wendt, M. Potkonjak, in *Security of IoT systems: design challenges and opportunities*. 2014 IEEE/ACM International Conference on Computer-Aided Design (ICCAD), (2014), pp. 417–423
14. B. Halak, M. Zwolinski and M. S. Mispan, *Overview of PUF-based hardware security solutions for the internet of things*, 2016 IEEE 59th International Midwest Symposium on Circuits and Systems (MWSCAS), Abu Dhabi, 2016, pp. 1–4. doi: 10.1109/MWSCAS.2016.7870046
15. H.M.Y. Gao, D. Abbott, S.F. Al-Sarawi, PUF sensor: exploiting PUF unreliability for secure wireless sensing. IEEE Trans. Circuits Syst. I Regul. Pap. **64**, 2532–2543 (2017)
16. M. Majzoobi, M. Rostami, F. Koushanfar, D. S. Wallach, S. Devadas, in *Slender PUF protocol: a lightweight, robust, and secure authentication by substring matching*. 2012 IEEE Symposium on Security and Privacy Workshops, (2012), pp. 33–44
17. Y. Gao, G. Li, H. Ma, S. F. Al-Sarawi, O. Kavehei, D. Abbott, et al., in *Obfuscated challenge-response: a secure lightweight authentication mechanism for PUF-based pervasive*

devices, 2016 IEEE International Conference on Pervasive Computing and Communication Workshops (PerCom Workshops), (2016), pp. 1–6

18. R. Plaga, F. Koob, in *A formal definition and a new security mechanism of physical unclonable functions*. Presented at the Proceedings of the 16th International GI/ITG Conference on Measurement, Modelling, and Evaluation of Computing Systems and Dependability and Fault Tolerance, (Kaiserslautern, Germany, 2012)

19. C. Hazay, Y. Lindell, Constructions of truly practical secure protocols using standardsmartcards. Presented at the Proceedings of the 15th ACM Conference on Computer and Communications Security, (Alexandria, Virginia, USA, 2008)

20. R. Canetti, Universally compostable security: a new paradigm for cryptographic protocols. Presented at the Proceedings of the 42nd IEEE Symposium on Foundations of Computer Science, 2001

21. J.M.-Q.D. Hofheinz, D. Unruh, in Universally compostable zero-knowledge arguments and commitments from signature cards. 5th Central European Conference on Cryptology, (2005)

22. S. Goldwasser, Y. T. Kalai, G. N. Rothblum, in *One-Time Programs*, ed. by D. Wagner. Proceedings on Advances in Cryptology—CRYPTO 2008: 28th Annual International Cryptology Conference, Santa Barbara, CA, USA, August 17–21, 2008 (Berlin: Springer, 2008), pp. 39–56

23. U. Rührmair, in *Oblivious Transfer Based on Physical Unclonable Functions*, ed. by A. Acquisti, S.W. Smith, A.-R. Sadeghi. Proceedings on Trust and Trustworthy Computing: Third International Conference, TRUST 2010, Berlin, Germany, June 21–23, 2010 (Berlin: Springer, 2010), pp. 430–440

24. M.v.D. in System and method of reliable forward secret key sharing with physical random functions, US Patent, 2004

25. C. Brzuska, M. Fischlin, H. Schröder, S. Katzenbeisser, in *Physically Uncloneable Functions in the Universal Composition Framework*, ed. by P. Rogaway. Advances in Cryptology—CRYPTO 2011: 31st Annual Cryptology Conference Santa Barbara, CA, USA, August 14–18, 2011 (Berlin: Springer, 2011), pp. 51–70

26. M v.D.a.U. Ruhrmair, Physical unclonable functions in cryptographic protocols: security proofs and impossibility results

27. M.O. Rabin, *How to exchange secrets with oblivious transfer*. Harvard University (1981)

28. S. Even, O. Goldreich, A. Lempel, A randomized protocol for signing contracts. Commun. ACM **28**, 637–647 (1985)

29. C.-K. Chu, W.-G. Tzeng, in *Efficient k-Out-of-n Oblivious Transfer Schemes with Adaptive and Non-adaptive Queries*, ed. by S. Vaudenay. Proceedings on Public Key Cryptography—PKC 2005: 8th International Workshop on Theory and Practice in Public Key Cryptography, Les Diablerets, Switzerland, January 23–26, 2005 (Berlin: Springer, 2005), pp. 172–183

30. M. Backes, A. Kate, A. Patra, in *Computational Verifiable Secret Sharing Revisited*, ed. by D. H. Lee and X. Wang. Proceedings on Advances in Cryptology—ASIACRYPT 2011: 17th International Conference on the Theory and Application of Cryptology and Information Security, Seoul, South Korea, December 4–8, 2011 (Berlin: Springer, 2011), pp. 590–609

31. T. Eccles, B. Halak, A secure and private billing protocol for smart metering. IACR Cryptology ePrint Arch., 654 (2017)

32. T. Eccles, B. Halak, Performance analysis of secure and private billing protocols for smart metering. Cryptography **1**, 20 (2017)

33. J.P. Carmo, J.H. Correia, in RF microsystems for wireless sensors networks. International Conference on Design & Technology of Integrated Systems in Nanoscal Era (2009), pp. 52–57

34. H.S. Kim, S.-M. Kang, K.-J. Park, C.-W. Baek, J.-S. Park, Power management circuit for wireless ubiquitous sensor nodes powered by scavenged energy. Electron. Lett. **45**, 373–374 (2009)

35. H. Liu, L. Cheng, D. Li, Design of smart nodes for RFID wireless sensor networks. Int. Workshop Educ. Technol. Comput. Sci. **2**, 132–136 (2009)

36. M. Conti, in *Secure Wireless Sensor Networks: Threats and Solutions.* Springer Publishing Company, Incorporated, 2015
37. T.F.E. Diehl, in *Copy watermark: closing the analog hole.* Proceedings of IEEE International Conference on Consumer Electronics (2003), pp. 52–53
38. Y.G.H. Ma, O. Kavehei, D. C. Ranasinghe, in *A PUF sensor: Securing physical measurements.* IEEE International Conference on Pervasive Computing and Communications Workshops (PerCom Workshops), (Kona, HI, 2017), pp. 648–653
39. E.G.a.R.K.K. Rosenfeld, in Sensor physical unclonable functions. IEEE International Symposium on Hardware-Oriented Security and Trust (HOST) (Anaheim, CA), pp. 112–117
40. U. Guin, K. Huang, D. DiMase, J.M. Carulli, M. Tehranipoor, Y. Makris, Counterfeit integrated circuits: a rising threat in the global semiconductor supply chain. Proc. IEEE **102**, 1207–1228 (2014)
41. A. Yousra, K. Farinaz, P. Miodrag, Remote activation of ICs for piracy prevention and digital right management. IEEE/ACM Int. Conf. Comput. Aided Design **2007**, 674–677 (2007)
42. P. Tuyls, L. Batina, in *RFID-Tags for Anti-counterfeiting,* ed. by D. Pointcheval. Proceedings on Topics in Cryptology—CT-RSA 2006: The Cryptographers' Track at the RSA Conference 2006, San Jose, CA, USA, February 13–17, 2005 (Berlin: Springer, 2006), pp. 115–131

Appendix A

A.1 System Verilog Description of a PUF Arbiter

The design described in this section is based on the PUF arbiter architecture shown in Chap. 2, Fig. 2.8. An exemplar description of System Verilog of this design is provided in Fig. A.1 flowed by the description of the design of NAND latch in Fig. A.2.

© Springer International Publishing AG, part of Springer Nature 2018
B. Halak, *Physically Unclonable Functions*,
https://doi.org/10.1007/978-3-319-76804-5

```
-----------------------------------------------------------------------

module ArbiterPUF        #(parameter n =32 )       //n indicate the number of
stages

                                        (output logic response,

                                        input logic [n-1:0] challenge,
                                        input logic pulse);

logic [n-1:0] I0, I1;


logic [n-1:0] O0, O1;


//Instantiate Switches

Switch s[n-1:0] ( .c(challenge),    .I0(I0), .I1(I1),  .O0(O0), .O1(O1));


//Assign pulse

assign {I0[0], I1[0]} = {pulse, pulse};

//Assign internal signals

assign {I0[n-1:1], I1[n-1:1]} = {O0[n-2:0], O1[n-2:0]};

//Instantiate LATCH

NAND_LATCH n0 ( .S(O0[n-1]),

                                .R(O1[n-1]),

                                .Q(response));

endmodule

-----------------------------------------------------------------------
```

Figure A.1 Arbiter PUF description in System Verilog

```
---------------------------------------------------------------------

module NAND_LATCH(        input logic S, R, output logic Q);

logic Q_N;

nand n0(Q_N, S, Q);

nand n1(Q, R, Q_N);

endmodule

---------------------------------------------------------------------
```

Figure A.2 NAND_Latch description in System Verilog

Appendix B

B.1 Introduction

It is useful in some cases to be able to model the architecture of a PUF design in high-level language. Such model can be used for early exploration of the design characteristics such as resilience to machine learning attacks. This appendix explains how to model an ideal arbiter PUF using a MATLAB and how to evaluate PUF performance such as uniqueness and uniformity.

B.2 MATLAB Modelling of a 16-bit Arbiter PUF

The following script creates an ideal instance of 16-bit arbiter PUF. This model is constructed using uniformly distributed random numbers as the delay parameters for switching components.

© Springer International Publishing AG, part of Springer Nature 2018   233
B. Halak, *Physically Unclonable Functions*,
https://doi.org/10.1007/978-3-319-76804-5

Save as > TopLevel_ArbiterPUF.m

```matlab
%Clear the command window and workspace
clc;
clear;


%--------------------Start of User Input-------------


%n-bit Arbiter-PUF
n=16;


%# of Arbiter-PUF instances
inst=1;


%Read the challenge
c = dlmread('challenge.csv');


%---------------------End of User Input--------------


%Generate delay parameters for Arbiter-PUF
[dir,cro]=delay_parameter_generation(n,inst);


%Configure the Arbiter-PUF function with the generated delay parameters
%And apply arbitrary challenges onto Arbiter-PUF model
for i=1:inst
array_str=num2str((arbiter_puf(c,dir(i,:),cro(i,:)))');
array_str(isspace(array_str)) = '';
response_str{i}=array_str;
end


%Write the corresponding responses to a .dat file
fid=fopen('Response.dat','w');
for j=1:size(response_str,2)
fprintf(fid,'%s\r\n',response_str{j});
end
```

Several functions are needed to run the above script successfully, as below:

Save as > delay_parameter_generation.m

```
function [dir,cro]=delay_parameter_generation(n,inst)
%For reproduciblity, could be any number
rng(492);
%Generate random challenges to verify the delay parameters
count=0;
sample=1000;
while count~=sample
challenge=randi([0 1],sample,n);
challenge_str = strtrim(cellstr(num2str(challenge)));
count=size(unique(challenge_str),1);
end
rand_C=challenge;
%Get the size
[n_C, n_bit] = size(rand_C);


%Pre allocation
%res = zeros(n_C,1);


%Generation of the delay parameters are based on 1000 random challenges
%Build an ideal arbiter-puf with uniformity about 0.45~0.55
u = 0;
for j=1:inst
while(u>0.55 || u<0.45)
    %--Randomly generate two vectors of delay difference
    dir_path = rand(1, n_bit) - 0.5;   %Direct path
    cro_path = rand(1, n_bit) - 0.5;   %Cross path


    %--Apply delay parameters onto an Arbiter-PUF function
    res = arbiter_puf(rand_C,dir_path,cro_path);


    %--Calculate uniformity
    u = sum(res)/length(res);
    fprintf('Uniformity:\t%6.2f\n', u);
end


%Store delay parameter in array
dir(j,:)=dir_path;
cro(j,:)=cro_path;
```

```
%Reset u
u=0;
end
end
```

Save as > arbiter_puf.m

```
function [r] = arbiter_puf(c,dir_param,cro_param)


dir_path = dir_param;


cro_path = cro_param;

%Get the number of challenges
n_C = size(c,1);

%Pre allocation
r = zeros(n_C,1);

for i = 1:n_C
    %--Feature transformation
    c_ft = arbiterFT(c(i,:));

    %--Inverted challenges
    c_inv = abs(c(i,:)-1);

    %--Delay path
    bot_delay = (dir_path.*c_inv + cro_path.*c(i,:)) * c_ft';


    r(i) = signNum(bot_delay);
end
end
```

```
Save as > arbiterFT.m
```

```
function [X_tf] = arbiterFT(X_org)


%Number of bits
k = size(X_org, 2);


%Pre allocation
X_tf = ones(size(X_org));


%Feature transformation
X_tf(:,k) = (1-2.*X_org(:,k));
for i = k-1:-1:1
    X_tf(:,i) = X_tf(:,i+1).*(1-2.*X_org(:,i));
end
```

```
Save as > signNum.m
```

```
function [s] = signNum(i)
    if(i>=0)
        s = 1;
    else
        s = 0;
    end
end
```

To create multiple instances (e.g. 30) of the design, one can set the parameter inst=30 in TopLevel_ArbiterPUF.m script. This can help emulate the silicon implementations of this PUF on different devices.

B.3 A MATLAB Script for Evaluating the Uniqueness of PUF

Uniqueness is the measure of the ability of one PUF instance to be uniquely distinguished from a group of PUFs of a similar type. Ideally, the value of uniqueness is 50%. The Hamming distance (HD) is used to evaluate the uniqueness performance and is called the 'Inter-chip HD'. If two chips, i and j ($i \neq j$), have n-bit responses, $R_i(n)$ and $R_j(n)$, respectively, for the challenge C, the average inter-chip HD among k chips is defined as

$$HD_{INTER} = \frac{2}{k(k-1)} \sum_{i=1}^{k-1} \sum_{j=i+1}^{k} \frac{HD(R_i(n), R_j(n))}{n} \times 100\%$$

```
function [uniquenessValue] = uniqueness(responseSet)

%responseSet sets of PUF responses, each row contains response bits for
each PUF instance.


k = size(responseSet, 1);      %The number of PUFs

n = size(responseSet, 2);      %The number of response bits

total_HD=0;

for i=1:k-1

    for j=i+1:k

        total_HD = total_HD + sum(abs(responseSet(i,:)- responseSet(j,:)));

    end

end

uniquenessValue = (2*sum)/(n*k*(k-1))*100;
```

B.4   MATLAB Scripts for Evaluating the Reliability of PUF

The reliability of the PUF determines how reliable the response of PUF given the same challenge at different ambient temperatures and/or voltage supply fluctuations. The Hamming distance (HD) is used to evaluate the reliability performance and is called the 'Intra-chip HD'. If a single chip, represented as i, has n-bit reference response $Ri(n)$ from the chip i at normal operating conditions (at room temperature using the normal supply voltage) and the same n-bit response obtained at different conditions $R'i\ (n)$, respectively, for the challenge C, the average intra-chip HD for k samples/chips is defined as

$$HD_{INTRA} = \frac{1}{k}\sum_{i=1}^{k}\frac{HD\big(R_i(n), R_i'(n)\big)}{n} \times 100\%$$

From the intra-chip HD value, the reliability of a PUF can be defined as

$$reliability = 100\% - HD_{INTRA}$$

```
function [reliabilityValue] = reliability(responseSetA, responseSetB)

% Size of responseSetA must match responseSetB

if(size(responseSetA,1)~=size(responseSetB,1) ||
size(responseSetA,2)~=size(responseSetB,2))

    disp('Error: Sample size must be the same!!');

    return;

else

    n = size(responseSetA,2);          %Number of bits per response

    k = size(responseSetA,1);          %Number of samples

end


%Compute HD for each row

total_HD = 0;

for i = 1:k

    total_HD = total_HD + sum(abs(responseSetA(i,:)- responseSetB(i,:)));

end

HD_intra = (total_HD/((k)*n))*100;

reliabilityValue = 100-HD_intra;
```

Appendix C

C.1 MATLAB Code for a PUF Modelling Attack using Support Vector Machine Algorithm

This is an exemplar script which can be used for developing support vector machine modelling attacks on an arbiter PUF. It uses the feature transformation function `arbiterFT` below.

© Springer International Publishing AG, part of Springer Nature 2018
B. Halak, *Physically Unclonable Functions*,
https://doi.org/10.1007/978-3-319-76804-5

Save as > `arbiterFT.m`

```
function [X_tf] = arbiterFT(X_org)

%Number of bits
k = size(X_org, 2);

%Pre allocation
X_tf = ones(size(X_org));

%Feature transformation
X_tf(:,k) = (1-2.*X_org(:,k));
for i = k-1:-1:1
    X_tf(:,i) = X_tf(:,i+1).*(1-2.*X_org(:,i));
end
```

Save as > `SVM_attack.m`

```
%Clear the command window and workspace

Clc;
Clear;

%------------------------start-of-user-input-----------------%
%Filename that contains challenge-response pairs (CRPs)
%-----------------------------------------------------------%

filename='CRPs.csv';

%-----------------------------------------------------------%
%Define the size of the training and test set
%-----------------------------------------------------------%
n_CRPs = 15000;          % training_size
n_end = 1500;            % test_size


%-----------------------------------------------------------%
%Define the number of bit per challenge
%-----------------------------------------------------------%

n_bit = 16;              % n-bit Arbiter-PUF



%-----------------------------------------------------------%
%Define the step size  for the incremental training process
%-----------------------------------------------------------%

step = 500;
inner_iterator = 1;

%------------------------end-of-user-input----------------%

%Read the data set
M = dlmread('CRPs.csv');

%-----------------------------------------------------------%
%Define the challenge from data set
%-----------------------------------------------------------%
```

```
X_org = M(:,1:n_bit);    % Challenges

%------------------------------------------------------------------%
%Define the responses from data set
%------------------------------------------------------------------%

y_org = M(:,n_bit+1);    % Responses

%------------------------------------------------------------------%
% Transform the challenge input into feature vector using Feature
Transformation function arbiterFT defined below
%------------------------------------------------------------------%

X_org = arbiterFT(X_org);


%------------------------------------------------------------------%
% Perform SVM Attack
%------------------------------------------------------------------%

%----Pre defined variable
accuracyIndex = 0;
traningNum = 0;
accuracyPre = 0.5;
accuracyPre_inner = zeros(1, inner_iterator);

%----Pogress calculation
totalLoop = ceil(n_end/step)*inner_iterator;
fprintf('Expected Work done...%%%6.2f', 0);

%----Plot the figure of the prediction accuracy
% plot(traningNum, accuracyPre, 'x');
axis([0 1 0 1]);
%title('Modeling Build Attack based on Machine Learning Methods');
xlabel('Number of CRPs used for training');
ylabel('Prediction accuracy');

for loop = 1:step:n_end
    for i = 1:inner_iterator
        %Choose CRPs for prediction
        %----Collect CRPs

        X = X_org;
        y = y_org;

        X_pre = X(loop+1:end, :);
        y_ans = y(loop+1:end);
        %y_ans_svm = y_svm(index);
        %----Delete chosen CRPs from orginal CRPs
        X(loop+1:end, :) = [];
        y(loop+1:end) = [];

        %Train the SVM Classifier


        puf = fitcsvm(X,y,'KernelFunction','linear');
```

```
%-------------------------------------------------------------%
        Evaluate the prediction accuracy
%-------------------------------------------------------------%

            [y_pre_SVM,~] = predict(puf,X_pre);
    %---Calculate one sample of accuracy
    accuracyPre_inner_SVM(i) = responseAccuracy(y_pre_SVM, y_ans);
        %=============================================================

            %----Collect data sets of prediction accuracy
    accuracyIndex = accuracyIndex + 1;
    traningNum(accuracyIndex) = loop;
    accuracyPre_SVM(accuracyIndex) = max(accuracyPre_inner_SVM);

    %----Update the figure
    hold on
    plot(traningNum, accuracyPre_SVM, 's');
    axis([0 max(traningNum) 0 1]);
    legend('SVM', 'Location', 'southeast');
    drawnow
end
nextline();
```

C.2. MATLAB Code for a PUF Modelling Attack using an Artificial Neural Network Algorithm

This is an exemplar script which can be used for developing artificial neural network algorithm modelling attacks on a PUF. It does not use feature transformation.

```
Save as > ANN_attack.m
%Clear the command window and workspace
clc;
clear;
%-----------------------start-of-user-input-----------------%
%Filename that contains challenge-response pairs (CRPs)
%Data in the file must be in the format <challenge><response>
%----------------------------------------------------------%


filename='CRPs.csv';


%----------------------------------------------------------%
%Choose ANN Training algorithm: resilient backpropagation
%----------------------------------------------------------%


training={'trainrp'};


%----------------------------------------------------------%
%Define the size of the training and test set
%----------------------------------------------------------%


test_size=2000;
training_size=18000;


%----------------------------------------------------------%
%Define the step size  for the incremental training process
%----------------------------------------------------------%


step_size = 500;


%----------------------------------------------------------%
%Define the number of bit per challenge
%----------------------------------------------------------%
bit_C=32;


%----------------------------------------------------------%
%Define the number of bit per response
%----------------------------------------------------------%


bit_R=1;


%-----------------------end-of-user-input-----------------%

%Read the data set
M = dlmread(filename);
[row_M,col_M]=size(M);


%----------------------------------------------------------%
%Define the challenge from data set

%----------------------------------------------------------%


M_challenge=M(:,1:bit_C);
M_challenge_T=M_challenge';
```

```
%------------------------------------------------------------------%
%Define the responses from data set
%------------------------------------------------------------------%

M_response=M(:,(bit_C+1));
M_response_T=M_response'

%------------------------------------------------------------------%
%Test set
%------------------------------------------------------------------%
test_challenge=M_challenge_T(:,(row_M-test_size+1):row_M);
test_response=M_response_T(:,(row_M-test_size+1):row_M);

%------------------------------------------------------------------%
 %Number of iteration
%------------------------------------------------------------------%

iter=(training_size)/step_size;
%------------------------------------------------------------------%
%Initialize array
%------------------------------------------------------------------%
c{:}=zeros;
cm{:}=zeros;

for i=1:size(training,2)

    for j=1:iter

        %Training set
        train_challenge=M_challenge_T(:,1:step_size*j);
        train_response=M_response_T(:,1:step_size*j);

        %For reproducibility - fix the seed
        setdemorandstream(491218382);

        %Feed forward neural networks
        net = feedforwardnet([32],training{i});

        %Training parameters
        net.trainParam.epochs = 1000000;     % Default value is 1000
        net.trainParam.max_fail = 15;        % Default value is 6
        %Default values
        net.trainParam.goal = 0.0;
        net.trainParam.min_grad = 1e-5;


%------------------------------------------------------------------%
%Initializes the weights and biases of the network
%Can also skip this step and go directly to train the network %train
command will automatically configure the network
%------------------------------------------------------------------%



        net = configure(net,train_challenge,train_response);

%------------------------------------------------------------------%
```

```
%Divide the training set into training and validation
        %to avoid overfitting
%----------------------------------------------------------------%
        net.divideParam.trainRatio = 90/100;
        net.divideParam.valRatio = 10/100;
        net.divideParam.testRatio = 0/100;
%----------------------------------------------------------------%
%Building the feed forward network - use parallel computing
%----------------------------------------------------------------%


        [net,tr] =
train(net,train_challenge,train_response,'useParallel','yes','showResources
','yes');


%----------------------------------------------------------------%
%Testing the network
%----------------------------------------------------------------%
        predicted_response =
net(test_challenge,'useParallel','yes','showResources','yes');
        testIndices = vec2ind(predicted_response);

        %Measure the performance of the network
        [c{j},cm{j}] = confusion(test_response,predicted_response);
        train_size(:,j)=size(tr.trainInd,2);
        val_size(:,j)=size(tr.valInd,2);
        cc(:,j) = 100*(1-c{j});


    end
end
%----------------------------------------------------------------%
%Plotting the results the number of training data
%vs. prediction accuracy
%----------------------------------------------------------------%
plot((train_size+val_size),cc,'c*');
ylim([0 100]);
title(My title);
xlabel('# X ');
ylabel('Y');
```

Index

© Springer International Publishing AG, part of Springer Nature 2018
B. Halak, *Physically Unclonable Functions*,
https://doi.org/10.1007/978-3-319-76804-5

Printed in the United States
By Bookmasters